OBJECT LESSONS

A book series about the hidden lives of ordinary things.

Series Editors:

Ian Bogost and Christopher Schaberg

Advisory Board:

Sara Ahmed, Jane Bennett, Jeffrey Jerome Cohen,
Johanna Drucker, Raiford Guins, Graham Harman,
renée hoogland, Pam Houston, Eileen Joy, Douglas
Kahn, Daniel Miller, Esther Milne, Timothy Morton,
Kathleen Stewart, Nigel Thrift, Rob Walker, Michele White.

In association with

LOYOLA
UNIVERSITY
NEW ORLEANS

Georgia
Tech

Center for
Media Studies

D0674186

BOOKS IN THE SERIES

earth

JEFFREY JEROME COHEN
LINDA T. ELKINS-TANTON

Bloomsbury Academic
An imprint of Bloomsbury Publishing Inc

B L O O M S B U R Y
NEW YORK · LONDON · OXFORD · NEW DELHI · SYDNEY

Bloomsbury Academic

An imprint of Bloomsbury Publishing Inc

1385 Broadway	50 Bedford Square
New York	London
NY 10018	WC1B 3DP
USA	UK

www.bloomsbury.com

**BLOOMSBURY and the Diana logo are trademarks of
Bloomsbury Publishing Plc**

First published 2017

Library of Congress Cataloging-in-Publication Data
Names: Cohen, Jeffrey Jerome. | Elkins-Tanton, Linda T.
Title: Earth / Jeffrey Jerome Cohen and Linda T. Elkins-Tanton.
Description: New York : Bloomsbury Academic, 2017. | Includes bibliographical references and index.
Identifiers: LCCN 2016025987 (print) | LCCN 2016026876 (ebook) | ISBN 9781501317910 (pbk. : alk. paper) | ISBN 9781501317927 (ePub) | ISBN 9781501317934 (ePDF)
Subjects: LCSH: Earth (Planet)–Popular works.
Classification: LCC QB631.2 .C64 2017 (print) | LCC QB631.2 (ebook) | DDC 550–dc23
LC record available at https://lccn.loc.gov/2016025987

ISBN: PB: 978-1-5013-1791-0
ePub: 978-1-5013-1792-7
ePDF: 978-1-5013-1793-4

Series: Object Lessons

Cover design: Alice Marwick

Typeset by Deanta Global Publishing Services, Chennai, India
Printed and bound in the United States of America

CONTENTS

1 PROLOGUE: GENESIS

In the spring of 2012 the two of us were invited to deliver a co-plenary at the BABEL Working Group Biennial Conference in Boston. Our common denominator was Arthur Bahr, a medievalist at MIT, who knew that we both had a strange passion for rocks. We are also both obsessed with fundamental, maybe unanswerable questions, about nature, humanness, and our place in the universe—as well as extinction, communication, the complexities of life, oscillations of scale, and the vastness of time. The differences in our disciplinary training attracted the conference organizers: one of us is a planetary scientist who at that time headed the Department of Terrestrial Magnetism of the Carnegie Institution for Science, the other a professor of English who directs a medieval and early modern studies institute. The organizers hoped we might stage a conversation across the natural sciences and the humanities, fields that too seldom collaborate. We admit that we were intimidated by each other and worried that we would have no common language. We hesitated to meet, even though we both lived at that time in the same city. A lunch in a Washington, DC sushi bar quickly dispelled those anxieties—so much so that we decided our conference plenary would be performed without a safety net. We prepared individual five-minute introductory statements about stakes, approach, and method. We then sat on a stage and conversed before the audience, without a script but full of hope. We inadvertently posed to each other profoundly difficult queries that get at the heart of disciplinary differences: about the Big Question a researcher in the humanities frames his inquiry around in the hope of resolving (a humanities researcher proliferates questions, he does not answer them!), about the role of beauty in scientific research as motivator and persuasive element (to a planetary scientist beauty is personal and therefore a threat to her data and interpretation!). We loved these surprising questions that we had never thought to ask ourselves, even as the audience sometimes squirmed, sometimes cheered. This book is our attempt to expand upon a conversation that began with Arthur Bahr and the conference in Boston and has yet to conclude.

Earth is faithful to the modes in which it was composed. The letters that we sent each other are not fictions. Though revised for coherence and to provide a sense of fullness and completion, the various transcripts, social media updates, and instant messages are the actual technologies and genres through which the book was written, not a literary conceit. If such devices and framings seem at times precious and disorienting, that may well be because our subject is at times precious and disorienting. We believe that Earth requires multiple modes of approach to convey both care and stakes. To be honest, we also used Twitter and Skype and face-to-face conversations, but we realize there are limits to readerly patience. We hope that what we are presenting conveys some small sense of the wide world within which this book was written. We hope that the book invites you into the conversation, as partner and as future.

2 ORBIT

Earth is a home, a limit, and a recurring challenge.

Humans have long struggled with their desire to view the Earth from its outside—as if we could depart the only dwelling we have ever possessed, the place of our birth, and turn back to see that world as a whole. We imagine such a view to be radiant, a revelation, and forget how much it obscures. Names for this planetary home have been many, and ways of describing its shape and function are as diverse as the human cultures that imagine them. This book focuses upon a long Western tradition of imagining the Earth as an encompassed sphere, from the classical and medieval *orbis terraram* (a circle of lands which when viewed from above resolves into a globe) to a spaceship or Blue Marble. Only recently has this singular Earth been reduced to one celestial body among billions in a universe indifferent to its splendor and unconvinced of its uniqueness. For a long time we thought the universe revolved around our globe. We now speak of a Copernican Revolution that jarred us away from this anthropocentricity. Yet we remain in some ways deeply earthbound.

The desire to encompass the Earth through a vision that offers a discrete object, a total picture viewable from its exterior, may be as old as the realization that dreams free us to wander, to create new relations to what we imagine we've left behind. Through technology and the imagination we attempt to escape the planet's gravity to attain a comprehensive perspective, looking down from far above, a vision of the Earth denied from its expansive surface. Girdled first by wooden ships and now by urban lights, airplanes, and the Internet, the Earth seems to have shrunk into Google Earth, a tamed and domesticated thing, a human commodity rather than a luminous celestial body that defies full knowing. But the image of the Earth as spaceship or marble or *orbis terrarum* or dwindled product of the Anthropocene is still framed from a human point of view. The image leaves much to shadow and obscurity, including the interior. It is easier to reach Mars than our own planet's core. The Earth is too large, too old, too inaccessible to our senses for us to fully apprehend it all at any one time. We live within the limitations of our human selves, making it difficult for us to contend with global-scale issues and events, like space travel and climate change. We look at images of the Earth and see at once a cosmic globe and a planet we have altered: oceans, flora, fauna, and atmosphere have all felt our heavy hand. Creativity and imagination enable us to push against these limits and disjunctions though, and perhaps to see ourselves as having enough agency and breadth of vision to face such difficulties of vision, apprehension, and action.

FIGURE 2.1 Terrestrial hemisphere map from the 5-volume *Tenmon Zukai (Illustrated Explanation of Astronomy)*, dated 1689, by Iguchi Tsunemori. With permission from the Ana and James Melikian Collection.

Earth is a problem. Much of what follows is about the challenges of its boundaries, depths, and scale, offering repeated approaches that seem to present a total picture but not quite. Think of all those satellites circling the Earth right now, each trajectory yielding a particular view—perhaps from Antarctica "up" past Australia to pass over the North Pole, then downward by way of South America; or maybe an orbit that traces only the equator, Somalia eastward toward

Brazil. All the views together can be computer collated to add up to a blue-white orb in the darkness of space . . . but even such widely published, richly composite images leave out more of the Earth than they contain. Very few humans have ever beheld the Earth in its entirety. Yet as ancient texts attest we have long imagined attaining that perspective, whether in sleep or in death or as an astronaut. We begin this book by acknowledging that a desire to imagine ourselves looking back upon the only home we have ever known as if we were at its outside, gazing at Earth become object, recurs across history. Yearning to behold the planet from some point of view more comprehensive than the fragmented perspectives we possess living along its surface—to see snowy mountains, torrid deserts, and roiled seas resolve into a singular sphere—is an enduring aspiration, and a ceaseless prod to creativity. We imagine ourselves leaving our terrestrial habitation to perceive in one instant its vexing expansiveness, to gain a total picture—to grasp the Earth, and maybe even to love it.

If the Earth is a singular object, how do we observe it? How do we know it? How does knowing the Earth challenge what *knowing* means in the humanities? In the natural sciences? How do the stories humans have long been telling about the Earth inform and interact with the Earth we apprehend from our expanded technological capabilities? Why do we so ardently desire to be able to look at the Earth as if we no longer stood upon its rough surface? Can technology intensify how we feel about the Earth? How are recent spacecraft- and satellite-derived images of the Earth that represent the globe suspended in the immensity of space connected to an ages-long impulse to imagine what the Earth looks like when viewed from a great distance? Within such a perspective are we looking back or looking out? Why does the urge to explore drive us away from the only home we have had? Is Earth's gravity as metaphorical a force as it is physical, and if so can we ever escape that pull? Do we travel into space (through probes or through story) only to discover the past or the future of the Earth? Or to try to discover if we are alone?

The question of Earth and the question of life are intimately bound. Our imaginations are earthbound: we can only think about life within the parameters of what we know from Earth's biological flourishing. How can we look for kinds of life that we cannot imagine? Or to ask the same question another way, how might we extend our imaginations to begin to grasp realms that exceed us? Can thinking critically across the disciplines about Earth render us more aware of what those limits are? What we know from Earth's surface, the thin portion of the planet on which we dwell, does not necessarily assist us in thinking about the Earth's interior (vast in its range of temperature and pressure) or atmosphere (so much thinner and more vulnerable than we experience it to be). Strangely though, our surface-fettered experience serves us better for knowing other rocky planets than it does for knowing Earth itself. Earth's surface is far more similar to the surface of Mars than it is to the planet's own interior. The question of Earth is always a questioning of scale: can we understand the inhumanly deep, vast, tiny, slow, swift?

What technologies (including narrative) can we use to better apprehend a planet that does not exist within a frame of time or spatial extension familiar to us? For millennia humans have conceptualized the cosmos as revolving around the Earth. How do we leave the geocentric (that is, the anthropocentric) modeling of the universe behind? What do we lose when we do so, at last? What opens in its wake? What transpires when these long-told stories of the Earth meet more technologically enabled understandings of the planet? What happens to imagination? What unfolds at the confluence of humanities and planetary science?

So many questions. Before we begin to follow them, though, we want to emphasize some things that humans have in general long known about the Earth, a globe not always recognized as a planet but forever understood as a beginning and a home. The Earth is the center of *our* universe, even if not the center of *the* universe. And yet we have perpetually tried to rise above it, to imagine a perspective from the stars looking back: either through the eyes of gods or some other celestial dwellers, or through our own. The earliest astrolabe is Babylonian and suggests that mapping the stars is inseparable from populating them and imagining our Earth being gazed upon from a perspective that can see our home in its entirety, as an object. Star maps are works of the imagination and technology. They're machines that can guide ships. They also populate the cosmos and produce stories.

Humans have long realized the Earth is likely shaped like a sphere or ball—even when they depicted the Earth as a disk of land surrounded by a circle of sky and sea. Greek knowledge of the shape of the planet is often dismissed as rooted in philosophy or religious belief rather than observation, so that if the Earth was held to be a sphere it was because that shape is perfect, rather than because practical knowledge deriving from hypothesis and measurement suggests that fact. Yet in the sixth century BCE Pythagoras argued that the entire Earth had of necessity to be round, and by the fourth, Aristotle used evidence like the changing position of constellations relative to a traveler headed south to prove its spherical shape. The mathematician and astronomer Ptolemy buttressed this idea by pointing out that when earthbound things like mountains or ships rise or sink below the horizon, they clearly convey that the surface of the planet curves. It would be very difficult to find Western writers from this time onward who held the Earth to be flat. The encyclopedist Isidore of Seville (seventh century), the English monk Bede (c. 673–735), the Italian poet Dante (1265–1321), and the unknown author of the *Book of Mandeville* (mid-fourteenth century) are among the numerous medieval authors who contemplated the Earth's spherical nature. The widely held but erroneous belief that Columbus sailed to the Americas in part to disprove that you could sail off the edge of the world says much about our prejudice against those who lived in the past and little about those who actually lived at historically distant eras. So far as we can tell, humans have at every time and in every culture been far smarter than they have generally been given credit for, and have consistently exerted their impulse to disbelieve, reimagine, and create anew (even if such acts of creation make use of materials close at hand, including recycled ones). The possibility that India might be reached by sailing west from Europe was first mentioned by the Roman geographer Strabo in the first century BCE. Earth's circumference has long been known to be astonishingly vast. At his research center at the library of Alexandria, Eratosthenes in 240 BCE calculated the size of the globe by noting the angle of shadows at home and in the more southern city of Syene at noon during the summer solstice. We are not exactly sure which of two possibilities the unit of measure Eratosthenes used indicates when he wrote that the Earth had a circumference of 5,000 *stadia*, but from what we can tell he was likely correct to within 1 percent of the actual size (though he may have been off by as much as 16 percent, impressive all the same). For centuries writers have imagined the Earth as viewed from outer space, even if "space" consisted for them of a substance called ether and rotated in concentric rings that conveyed planets and stars along with them. That the Earth circled the Sun was first suggested by Aristarchus of Samos (third century BCE), though the idea did not become widespread until a good deal after the Copernican Revolution posited a heliocentric universe (a model that still did not get things right: the universe possesses no pivot that it spins around, and the Sun is of no more importance than any other star). The Earth quickly became one planet among a multitude of other planets, a diminished home. Using Murano glass to fashion powerful telescopes, Galileo observed four moons rotating around Jupiter, the first celestial bodies that could be shown definitively not to circle the Earth. Galileo also noted the pockmarked surface of the Moon and realized that the heavens were not a place of perfect light and unchanging shape as some philosophers had held, but were filled with stories of alteration over time. Eighteenth-century geology is usually credited with establishing the inhumanly long duration of the Earth, pushing its age into the millions and then billions of years. Biblically based calculations were obviously much shorter (and are still embraced by some Christians): they generally estimate the age of the Earth at six to nine millennia. Astute readers of Genesis, medieval theologians were quick to point out however that day and night did not exist before the sun and moon were created on day four, and that no one could know how long a day might extend for an everlasting and omnipotent deity. Some held the Earth to be indeterminately old, and others thought it eternal. No matter what the age of the Earth was believed to be however, writers discerned in its archive of rocks, strata, and fossils stories of erosion, sedimentation, and catastrophe that might repeat. Earth's time is not human time, but human lives are nonetheless bound to an Earth they imperfectly comprehend.

We continue today an ancient tradition of asking questions about what our Earth does, what it was in the past, and where it is going. For millennia we were limited to the observations we could make as we walked around on its surface, however much we might have yearned for and imagined other points of view, both interior and exterior. We are making new progress in understanding our planet as technology has offered new kinds of observations. As a part of the almost infantile sense of omnipotence delivered by technology comes such rapid progress in understanding Earth systems that we cannot integrate and synthesize the knowledge fast enough.

Physical measurements give some of the most profound insights: the ability to measure the magnetic field recorded in the rocks of the ocean floor while they were cooling after first formation provided the data that proved we have plate tectonics. The rigid cold plates on the exterior of the Earth move too slowly to be perceived over a human lifetime, but the bands of differently magnetized rocks in the ocean floor showed a record of their formation over time, like stripes in a blanket as it rolls off the loom. This measurement required not just ocean-going ships towing magnetometers, but a civilization affluent enough to send them to sea just for scientific measurement. Most importantly, it required people with enough imagination that they could see the data and create in their minds the scenarios that would produce that data, and then figure out ways to test whether they were right. This kind of learning—scientific learning—takes a surprising amount of imagination and creativity.

Earthquakes are wondrous and also viscerally terrifying and the Earth has had them throughout its existence. They occur in greatest numbers at the joins between the Earth's plates, where the plates move relative to each other. We still cannot predict when they will happen. The strains between the plates build up slowly, at the rates at which the plates move (centimeters per year), and we can't yet model the complexity of the systems that determine when the strain is too much and the faults break and the violent waves from the breakage radiate out into the world. In some senses we are in a similar place with earthquakes as we have been since the dawn of humanity: they frighten us and kill us and we do not know when the next one is coming. The waves the breaking fault sends out into the world do more than cause flocks of birds to take off and buildings to fall down. Some waves travel only along the rock-air interface (surface waves, naturally enough), and others radiate outward in all directions and thus travel down through the Earth, bouncing off of, moving along, and passing through all the boundaries between layers in the Earth's interior, and emerging at the surface elsewhere on the planet, full of information about what they passed through on their way. Just like noninvasive medical scans, earthquakes provide the waves that allow us to do tomography on the Earth.

Back in the eighteenth century, estimates of the Earth's density showed that it had to have a very dense core, and that core must be made of iron, the most common dense planet-forming material. Using earthquake waves in the 1930s, Danish scientist Inge Lehmann discovered that our solid iron core has a liquid inner core. This is the only liquid part of the Earth's interior. All the rest—the outer metal core and the thick rocky mantle—are solid. With another leap of imagination followed by decades of exhaustive lab experiments and observations to prove it, we now understand that the rocky mantle of the Earth flows like a liquid very slowly over geological time, even though it is solid. The high pressure and heat allow crystals to deform under stress and create great currents of flow in the mantle, like super-slow-motion molasses.

What happens to our imaginations when these meticulously tested and groomed data come along and place facts about processes and ages, like granite mile markers, in our mental highway? We are spurred to greater creativity. Science does not progress without brave leaps of imagination. And then our observations carry us forward with even greater leaps: Every time we can make a new kind of observation about the natural world, we see things we had never imagined—the sulfur volcanoes on the moon Io; the Earth's liquid inner core; the fact that almost every star has planets orbiting it; bacteria living at temperatures above the boiling point of water, or in the cores of nuclear reactors, or many kilometers underground. And then our imaginations are prodded to new heights, as if we were being challenged to be more creative than the universe. How human, who could resist?

3 GROUND (WHY EARTH?)

Dear Jeffrey,

Recently I've been working on research that will help explain how the Earth got its water. Why do I care about that? Because one of the biggest and longest-standing questions humankind asks is, are we alone? As a physical scientist (and not a philosopher, or a poet, or a minister) I then think about what life requires for existence. The one common element to all life we have found so far is water. Thus, for a planet to be habitable, we think it has to have liquid water. Then . . . seeking to find the bite-size increment of scientific knowledge that we can achieve, I approach the many steps of planet-building to see if water can be obtained and retained through all of those steps.

Again, why do I care? Because (and finally we are at the heart of the matter) if planets like Earth obtain their water during the natural process of accreting rocky material together in the early solar system, and not through some later additions by chance, then all rocky planets everywhere in our universe are more likely to be wet, and therefore habitable. If instead having water relies on the chance of later processes, after the planet is formed, then calculating how many habitable rocky planets there are out there becomes far harder, and the likelihood of a given rocky planet being habitable also sinks.

Something we don't talk about as much in science is the arc of emotion we experience as we go through this thinking. By the time we are thinking (and hoping) that there's a good chance all rocky planets in the universe might once have had, or still might have, a water ocean, we are imagining seeing these planets, watching the early life emerge, hoping to travel to see it one day, or at least find a way to detect it from Earth. These emotions are at the heart of humanness. Joy at creation, intense curiosity to see, to travel, to explore. Great hope that there are others out there, and more joy at the thought of the complexity and endlessness of the universe.

Then comes the frustration and cosmic chill when the problem of scale enters in. We are not yet universal explorers, at least, not in our bodies. We look out at the universe and we devise ways to detect the interior of our Earth, to some depth, but when it comes to experiencing it with our persons, we are thwarted. We cannot yet exist in deep space. We cannot dig more than a few kilometers into our Earth, and we cannot go there in person: Too hot, too much pressure, no technology to allow it. Even exploring the tops of mountains or the ice caps requires more strength and determination than most of us have. Sometimes walking to the store even seems too much. What do you think about the relative rewards and disappointments of exploring physically versus exploring in our minds? What kinds of truths can they each reveal?

And back to this question of water: by measuring the water in rocky meteorites and asteroids and comparing it to water on the Earth, my colleagues have shown that the Earth's water matches these rocky building blocks, and not the water in comets (we used to be taught—maybe you were taught?—that Earth's water came from comets, delivered after the Earth was formed and cooled). With computer models I have shown that water can be retained by the Earth through even the unimaginably giant impacts that grew the Earth to its current size. And here's my favorite part: NASA missions to Mercury, to the Moon, and to Mars have shown that those bodies were not completely dried out by their formation, either. I'm now an evangelist for planets getting their water through their common formation processes, and not by later chance, and so rocky planets through the whole universe have a chance at water oceans, and therefore life.

—Lindy

Dear Lindy,

Your letter about plumbing the Earth for water and the hope of life arrived just as I was thinking about floods. You noted that we used to believe that Earth's oceans arrived from elsewhere—transported on comets that smashed the dry terrain and made life possible. I've been writing about the tales humans have been telling for a long time—at least four thousand years, nearly as long as we have been able to send such stories into the future—about water arriving from elsewhere, but in a different way: ancient narratives of terrestrial inundation, like the *Epic of Gilgamesh* and the biblical story of Noah's Flood. Water is complicated. In Genesis heaven and earth are created through a divine command, but water already seems to be there, an oceanic abyss that God's spirit moves across before creation really begins (Gen. 1:2). When God sends the deluge to purge the Earth, the unleashed water comes streaming from the vault of heaven as well as churning from the bottom of the seas. You wrote about the hope that water gives, since without H_2O no life we know can arise. Why then do we so often associate water with the scouring of the Earth, with wiping the planet clean of its ecosystems? I'm writing this letter on the tenth anniversary of Katrina, and am reminded how easily we accept the inevitability of floods that destroy the lives of fellow humans. Water is the precondition of life, and yet we resign ourselves to repeating its ancient role as conveyor of death. Sometimes these old stories are not doing us much good.

Look at the narratives we tell about global warming: the polar ice caps will melt; sea levels will rise; our cities will submerge; Earth will become a planet of the drowned. I have this theory that our dominant mode of framing climate change is through the Noah story, transporting myth into the narration of science. Even though we're supposed to be secular, we repeat an inherited apocalypse without thinking about what else that story conveys. Christian theology praises Noah for his obedience. Jewish and Islamic traditions can be more ambivalent. Midrash, for example, describes Noah as righteous only relative to others in his generation, but not in relation to someone like Abraham. When God declared the destruction of Sodom and Gomorrah in a flood of flame, Abraham asked if God would really destroy the innocent with the guilty. His challenge worked. Noah compliantly built the ark and left the world to drown. I have a hunch that a reason we are so resigned to climate change and losing land to water is that we have internalized Noah's acquiescence to catastrophe. A story about a selfish family becomes a story about the whole Earth. I also think we should be aware that when climate disasters strike, it's always the economically and environmentally disadvantaged who are left outside the ark or the gated community. The aftermath of Katrina taught us that lesson well.

So we are writing a book together about this object called Earth, and here we are in deep water. And maybe that's because the element is so intimate to Earth's identity as a planet: a reservoir of the stuff inhabits its interior, and three fourths of the surface of the globe is covered by liquid rather than land. Maybe "Earth" is a misnomer and we ought to have called the planet Water. Shifting scales means shifting between materialities, shifting back and forth between the organic and the inorganic, between a hope for life and threats of extinction.

—Jeffrey

Dear Jeffrey,

Here in the desert of Arizona, particularly on a record-breaking day like today when the temperature reached 117 °F, water is an especially fraught subject. The western states are all in a drought and we are all overusing our water. We know we have too little. The curve-billed thrashers and even the hummingbirds in our yard pant with their bills open, and drink from the pan of water we put out. We pass the unaffordably wet golf courses with sinking hearts. But even our little desert water supports a vast amount of life, not just people, but a thriving desert flora and fauna seemingly even more abundant than the water-loving communities in the forests of New England.

Arizona is far wetter than the Moon. Arizona is wetter than Mars. We look for life on Mars, though, hoping there could have been better circumstances in the past, or that there could be the hardiest of microorganisms somehow living in the freezing brines there today. Water is what sets our planet apart, in the end.

When you wrote that water is complicated I thought how right you are. We require it for life and we talk about how watery Earth is, and yet somehow we are running out. Water destroys and drowns and terrifies, as in both your mythical and storied floods and in the floods we learn about in the news, or more awfully, in our towns or in our streets. Water drips through our dreams and images just as it flows through our bodies and irrigates our gardens and streams through our forests. Water is as central to our bodies as it is to our psyches and all the life on our planet.

Scientists do not talk about beauty as much as we might. Beauty is at the center of all we do, though; it is the emotion and response that compels us to study the things we do. Mathematicians speak of beauty as elegance and simplicity. Those of us who study natural science are often hypnotized by the beauty of fluid flow, whether it be flame, or clouds, or lava, or water. In *Sensitive Chaos* Theodor Schwenk laments that humankind has lost touch with the spiritual nature of water, but I suspect such a thing is impossible.[1] We are just fish living on our blue Earth, worshiping its water every day.

As easy as it is to slide into the poetic, how wet is the Earth really? Earth is wetter than our neighboring planets, but far drier than the icy and ocean moons of the outer solar system. Earth's wetness is relative and it is a matter of perception. We think that Earth's interior is far drier than its surface. If our current interrogation of the interior is correctly interpreted, then we survive and swim because water has been expelled from the interior to the surface through volcanism, and water has not yet been stripped away from the atmosphere by the solar wind. We're surviving and swimming in a puddle-thin layer of wetness painted over the surface of a dry, hot rock, and shielded from the tearing solar wind by very little indeed.

Perhaps this is the answer to, why the Earth? We're living in an ephemeral pool that to our brief human lives looks like an infinite supply.

All my best from dry dry (relatively) Arizona—

—Lindy

Dear Lindy,

So I suggested that Earth might better be called Water, since its surface is so wet. Others have argued that we rechristen the planet Ocean. You make it clear that if we are going to give a watery designation to the globe it would need to be more humble: Puddle. The Earth's water is only seemingly plentiful and Earth's story contains more rock than sea. So why do we act as if the "ephemeral pool" of lingering water were an inexhaustible resource, cultivating golf resorts in the deserts of dry, dry (relatively) Arizona? Why are we so wasteful with elements that are not infinite? Maybe part of the problem is that Arizona does not seem to be part of the Earth.

What I mean is that Arizona and Washington, DC (where I write to you from, and where on this late summer morning the air for once is not swampy but crisp) often seem so far apart, so unconnected that—despite being states in the same nation—it is easy to forget that on a larger scale they are regions within connected systems of geology (they are both gliding atop the same tectonic plate, they are both part of a single lithosphere) and climate (the troposphere is also a shared puddle or ephemeral pool). Last night we had the most remarkable sunset, rose and deep crimson smudged with fiery orange. The intense colors were triggered by smoke particles in the atmosphere streaming from the Pacific Northwest, where massive forest fires have been nurtured by long drought. We board a plane to Portland this evening to bring our son to college. There is a health advisory in that city about breathing too much ashy air. Sprinklers watering golf courses in Phoenix and arid stretches of what had been lush forest are part of the same problem, an overlap of what we segregate into regions (climates, biomes, states). We are prone to wastefulness and pollution because nearby lawns, streams, and skies do not seem to extend very far, even when they do. Maybe we need some Earth thinking to jolt us to a scale at which the causes and effects of climate change are easier to apprehend.

In her book *Sense of Place and Sense of Planet*, Ursula K. Heise has argued that a love of the local seldom translates to regard for the global.[2] It used to be assumed that attachment to forests, streams, sand, stones near to home would inevitably stir a desire to cherish and preserve nature, to cease imperiling the Earth as a whole. Evidence suggests though that the first step to thinking globally is not necessarily passionate localism. The much-vaunted connection to nature has too often meant a glib affection for the beauty of pines or dunes, eagles or beavers, as if such attachments could overcome alienation, resourcism, and the outsourcing of waste. We remain alienated from the natural world . . . and we assume we know what we mean when we use that term, "natural world." What does it mean? Even if everything is connected, only some of those connections are going to be palpable at any given time, are going to spur us into action—into maybe ceasing to plant desert golf courses, but also ceasing to rely so heavily on fossil fuels across the planet. In her book Heise argues for an "eco-cosmopolitanism" that might shift perspective from (say) dry Arizona and humid

Washington, DC to a globe that is awash not in water but in human political and cultural activities with ruinous, worldwide consequences. Might planetary science have something to add to this humanities-inflected conversation? Is a sense of planet even enough? You wrote about the imagination as a spur to science, and now I am wondering if imagination is necessary or is it something—because it is so human, so limited, so local—that always gets in the way.

Yours from Washington, DC, as we prepare to board a plane to Portland (a good place to think about water and what it feels like to be drenched by ecological connectedness),

—Jeffrey

Dear Jeffrey,

Your photos of your family delivering your son to college are so sweet, and so poignant. Here is the moment that your brave and talented boy departs from the immediacy of the family; it's a true moment of passage. We all know that moments like these make up a life, each of our lives, fragment by fragment. And yet at least for me each one looms up frighteningly fresh at the time, unknowable until it has occurred.

There in your moment with your family I think I saw the totality of the issues we are grappling with. We humans are so limited in our understanding of scale. Today is what we feel and understand the most, and a little less yesterday or tomorrow. We plan for our families and ourselves into the months and years but they are increasingly hazy and our faith in them accordingly low.

We are hot and thirsty today, and so we drink and we swim in our pool, and we like the view of the plants and so we water them. We want to have a nice afternoon and we're used to golfing with our business partners, and so off we go. The time it will take to use all our water as a society is vague and fuzzy to us, seeming irrelevant in the moment of our afternoon.

And then there is length. Time is bad enough, but length, size, volume are also impossible. We have a good sense of the distances we can walk or bicycle, and a warped sense of driving. We cannot really comprehend planet-scale processes, cannot really comprehend the size of the Earth, the volume of its water, the depth of its atmosphere. Imagine our boundless oceans and heaven-high atmosphere being just a paper-thin ephemeral pond coating our vast rock of a planet! You and I write these words, but it's hard to digest them, hard to react with anything other than an amused snort. What do we do with this information?

This is why, I think, that passionate care about the local does not translate to the global. We cannot understand the scale, and we cannot connect emotionally to it. We can love the animals and plants we live among, and care for our neighbors, but it's hard to feel truly personal about places that are too far away, too big, too distant in space and also time. How can we care in a personal way about the future? We can care in an abstract way, and we can care in a guilty way, but it's hard to care in the same way we care for our sons on the day we drop them at college.

The challenge, then, is to take what I'll brazenly call leaps of faith. We need to imagine the future us as lovable and personal, and hopeful and active. We need to imagine the world we want to live in, and then we need to make that world. We need to allow our convictions to overcome our inabilities to contend with scale.

From this place of concern and oblique discussion of limited resources, I hope we will soar into the beauties and immensities of scale as we attempt to apprehend our Earth. As the Apollo astronauts allowed us to see for the first time, our planet is a place of exceptional beauty. We continue to try to see and understand it.

All my best to you and yours on this the week of my fiftieth birthday—I am halfway, I think, and perhaps this moment of passage will bring me new appreciation of time!

—Lindy

Dear Lindy,

A very happy birthday to you! Selfishly I hope that you are less than halfway because I want these letters to pass between us forever. Even after we have completed our book on Earth, won't it be time to grapple with planetary systems, galaxies, the universe? And how can we speak of so large a scale without also plumbing the microscopic? Next up in the Object Lessons series: the very small book we will compose together on elementary particles. Or the microbiome.

I am joking, of course—although serious when I say that thinking about the very large (like the entirety of the Earth, or the future in its long duration)—also requires contemplating the distant and the impossibly small: ephemeral and difficult things like elements we cannot see, lives that are not ours, places and times to which we are minutely connected and yet because they are not palpably present do not seem to matter so much as neighbors, local resources, all things close at hand. They do. That's what the leap of faith you describe requires: embracing the intimacy of our entanglement with unknowns, realizing that the purified water that tumbles from our faucet is only a briefly domesticated version of the substance we pollute with plastic in our depleted oceans, that we divert from plants and animals and humans who need its sustenance, that as a life-giving puddle encompassing much of the Earth is too quickly being dissipated. Strange to think that a planetary ethics should ambush us every time we swim in a pool, play some sports, hydrate thirsty houseplants. But that leap of faith is also a leap into philosophy and scale. I am not sure what the Earth as object—as total system, as a thing that can stand alone—means in and of itself.

Probably nothing. The Earth *doesn't* stand alone, and is not a solitary object, despite the still, quiet reverence of its Apollo images. My favorite of these cosmic photographs was taken in 1968 by William Anders and is now called *Earthrise*. A radiantly blue globe, half cloaked in shadow like a waxing moon, is suspended above dusty lunar terrain. And I do mean suspended: the Earth hangs in an utterly black sky, serene and alone. *Earthrise* has been described as the most reproduced photograph in history as well as an essential prod to the awakening of a world environmental consciousness. Yet the image deceives. Orbiting in Apollo 8 at considerable distance from the surface, crewman Anders witnessed the shadowy Earth appear to the side of the Moon (he was orienting himself to the lunar poles, so the Earth "rose" to his left; it was not "above" anything). When published, Anders's photograph was rotated ninety degrees to provide its now iconic perspective. The blue Earth that hovers in a black sky seems as if beheld from the ground, just as we might witness the slow ascent of the Moon over a terrestrial horizon, and that perspective switch, familiar and alien at once, gives the image emotional force. We are used to thinking of the Moon as a beautiful object that does things in the sky, but not the Earth (here acting as if it were a moon of someone else's planet). A small point, maybe, but one that suggests that the Earth as object is always coming into being, coming into force. In *Earthrise* the black background which frames its difficult immensity obscures some human labor— and maybe a human desire for a beautiful home that exists immune to our actions, unpolluted, eternal, always going to rise in that lunar sky.

Like the *Blue Marble* photograph to follow (Apollo 17, 1972), the *Earthrise* image yields no hint of the planet's incessant, dizzying motion. Earth is whizzing through space, whirling within a vortex we stabilize into peaceful concentric rings and call the solar system, but it's really the plaything of solar and galactic gravity and so many forces that you understand so much better than I do, Lindy. Every image of the Earth that makes the planet seem lonely and immobile is wrong. I imagine the solar system is more like a collection of particles in a hurricane, a vortex of orbs spinning through space, than a collection of bodies serenely rotating around a centering Sun.

Earth can never be alone—and I think that's what James Lovelock's Gaia Theory gets wrong. Lovelock imagines his Earth as a closed system, respiring, adjusting to change, its balance enduring if delicate. The planet Lovelock describes seems a complicated but single-celled creature. Humans are a potentially disruptive presence, maybe even a hostile virus that will after heedless proliferation exhaust their resources and perish. Yet no organism, not even a planet that functions as if a self-regulating organism, is hermitic. Or abidingly balanced. Every time an aurora flares along the poles, aren't we reminded of how precarious a membrane the atmosphere offers? Its green and purple is a gift of the Sun we spin around, a warm star that enables our lives, but the aurora is a reminder that it is a star that does not love us. The sun is a blazing celestial body that simply *is*. Sometimes its solar wind drenches us with charged particles that remind us of the harm it can do to organic life. Thank goodness for our atmosphere, and for Earth as Puddle of Life. Neither planet nor star will last forever, but the Leap of Faith we make reminds us that the future those after us will inherit should not been ruined at our hands just because we know that nothing lasts.

Those are gloomy thoughts, perhaps, to offer so soon after a significant birthday. I also recently turned fifty, and find it hard to fight my urge to go epochal when I write. OK I'm lying: fifty is teaching me to resist that urge I've always had a little better, because humans—especially scholars—have a way of thinking and writing so gravely that the Leap of Faith doesn't happen. We are too earthbound, too much enamored of the future as the promise of the planet's grave. If we are doomed at least we do not have to care about our own agency. If you read the *Epic of Gilgamesh* with its world-destroying flood, you realize quickly that humans have *always* loved their apocalypses. For the Leap of Faith to take flight—for us to see that thinking the Earth means thinking its expansive yet fragile life as a swift Puddle of Life—requires some levity, something that lifts us outside ourselves. The Leap of Faith requires a Leap of Beauty: a recognition in the moments we inhabit that something exists larger than ourselves and more fragile than we imagine that is worth preserving for people (and plants and animals and even minerals) that we do not know, into futures that stretch beyond what we can comprehend.

I am thinking of a video chat we had with our son last night. He is so happy in Oregon. He had just returned from camping on the slopes of Mount St. Helens, sleeping with some friends in bivvy bags rather than tents so that they could watch the night stars. The next morning Alex took a strenuous hike to behold the devastation of the 1980 eruption: landslides, heaps of ash, shattered trees, a lake newly formed within what had been forested slope. Ever since he was very young Alex has been fascinated by the place. When he was four I read him a children's book about Mount St. Helens erupting. He was entranced by the geologist who warned everyone that the mountain was going to explode, and the elderly man who insisted on remaining inside his trailer within sight of the peak because he had always lived in that home. A massive pile of ash buried the old man and the book did not say if his body or trailer were ever found. I was shocked when I read those words to Alex, who so far had been raised on tales that end with reunions, triumphs, and happy revelations. He cried. Then he became obsessed by how someone could choose to remain in place, remain at home, when a mountain is shaking and a geologist tells you to flee. He wanted the book read to him again and again. And as a student young in college he made his pilgrimage to the place to behold its beauty. He knows that Mount St. Helens's allure includes its inhuman force, its unpredictability, the precariousness of life along its slope. The mountain is sublime not because it is eternal but because it could shift and yaw at any moment. It holds a geological life that is part of the life of the Earth. That's why it should be sought to sleep upon, to wander, to hold in awe. But the volcano, like the Earth, is not a safe or enduring home.

Nothing lasts. The future is uncertain. All the more reason to shelter the places and times where we expend our short lives, ensuring that those who come after us have the same chance to make the Leaps of Faith and Beauty that we have been offered.

Let's revisit these letters when we both turn 100.

—Jeffrey

PS Today I was reading of Enceladus, a watery moon that orbits Saturn. Although covered in an ice sheet, the planet wobbles enough that scientists know that an ocean sloshes beneath that layer of freeze. The icy plumes it sends into space are gorgeous, as if Old Faithful could breach the atmosphere of Earth. If we are still around at the century mark and space travel has improved to the point where humans who use canes and walkers can travel the solar system easily, could we plan on staging a reading of this book atop that frozen sea? As you have taught me, it may have life within.

4 SCALE (BARRIERS TO UNDERSTANDING)

Dear Jeffrey,

Nothing may last, and nothing may be safe in the end, not even our earthly haven, but "nothing may last" is such a relative phrase. There are lifespans for all things on Earth and in Heaven, so to speak.

When I teach introductory geology I have a special lecture about scale. The scale of the universe is incomprehensible to us, and the scale of the galaxy, and the scale of the solar system, and even the scale of the Earth is hard for humans to understand. I talk about the outrageous length scales we need to grapple with: from the ultramicroscopic to the interplanetary (the Moon is about 380,000 kilometers from the Earth, but Pluto can be as far as 4.67 billion kilometers from Earth), to interstellar distances that get measured in light years.

Temperature is no less difficult to grasp. The coldest we have created in a lab is just a few billionths of one degree above absolute zero, when atomic movement ceases. And the hottest temperature created in a lab was 2 billion degrees

Celsius, at Sandia National Labs. That is 100 times hotter than the interior of the Sun. In a lab!

And then there is time. Time from any fraction of a second, as arbitrarily small a fraction as you choose, up to the age of our solar system (4.567 billion years) and the age of our universe, 13.82 billion years.

Where to start. What is a billion, really? We can say it and write it (1,000,000,000) and we can calculate it, but can we really understand it? Do we have an internal sense of how many is a billion? And while we are at it, what does it mean for the universe to have an age?

For hundreds of years geology and many aspects of astronomy and cosmology lagged behind biology and chemistry, largely because of problems of scale. We are happy thinking of lengths that relate to our body. We measure horses with hands, and fields with strides. When we get beyond a football field or a mile or so, then we start thinking in terms of how long it will take us to traverse it, not really how long the length is.

For temperature, we think the temperature of water freezing is chilly, and we can't tolerate anything near the boiling point of water. That's such a small fraction of the available temperature range of the universe that I'm not going to bother writing down the ridiculous decimal.

We live on a human scale. We comprehend, on a fundamental, gut level, minutes and hours and days. Years, a little bit. Decades are a bit fuzzy. With these limitations, how could we comprehend the millions of years it takes to make a rock, or the thousands of kilometers into the interior of our planet, or the slow, slow process of erosion? Watching films, with their fast-forward and slow-motion capabilities, have helped us conceptualize. But these are not subconscious comprehension. They are a level removed from intimacy. They are familiarity, but not understanding. They are sympathy, but not empathy. It's a mechanistic relationship.

Like humans, some natural processes have lifespans. A planet is born, and then it goes through billions of years of active life, and then it dies. The same for a star, or a galaxy. The same for a coastline or a continent. Where do these long, long lifetimes intersect with our own short lives? In some cases, natural processes speed up to a level that we can comprehend them. During the lifetime of a waterman on the Chesapeake Bay, rising sea level and sinking land level conspire to change landmarks and cause him or her to feel lost in what used to be a familiar landscape.

During the long active lives of planets other events occur at some average tempo. Big earthquakes happen every year. People remember what they are like, and though they are devastating, they are also familiar. Big volcanic eruptions happen every hundred years or even less often. How long does it take for a person to forget? For a community? If an event happens once in a lifetime, we will know to expect it, but we will be surprised. If an event happens less than once in a lifetime, then new generations will come to think that those events no longer happen. How much time needs to pass before people feel safe, before our unconscious minds think it can't happen here any more?

These are limitations of scale. And we are creatures of small scales, no matter our precocious historical records and scientific investigations. The data may say where Pluto is or that temperatures are slowly rising or that the Sun will eventually bloat to the point that the solar system ends, but our primitive primate brain stems say, that is not our experience. Our experience is here, and now, and we can only clearly remember a few years back.

The beauty comes as we make little steps of progress in understanding scale. When a perfect animation gives you a sense of how far away the Moon is for the first time, your hair stands up and a shiver goes down your spine—you've had a moment of personal evolution. I think that is what is addictive about learning, and especially about discovery. We humans are innate explorers and we constantly strive to find the new.

Jeffrey, what is the challenge of scale in the study of human history? Is a thousand years ago as unreachable as a hundred thousand kilometers away? When we leave our immediate sphere of scale, is everything equally remote?

With all my best on this philosophical Saturday afternoon, feeling so fortunate for the luxury of mental exploration.

—Lindy

Dear Lindy,

Your letter reaches me just before my own birthday. One revolution around the Sun almost finished, another about to begin. Four and a half billion solar spins completed by Earth and counting. I know a single year of my life is nothing on this geological scale. Yet as you wrote in your last letter, we live on a human scale, not a planetary one. I will therefore celebrate that small cycle's completion with some cake.

The last twelve months have teemed with activity and movement. I have traveled by plane to Victoria British Columbia, Maine, Vancouver (twice!), Manitoba, Atlanta, the Yucatán Peninsula, Kalamazoo, Geneva, London, New Zealand, Portland, Oregon, and Cambridge, UK. Academics get around! It is interesting to consider that the reason for all this movement has to do with a substance usually considered immobile. Much of my professional travel this year involved giving talks drawn from my recent book, *Stone: An Ecology of the Inhuman*. As a result of this book on heavy matter I spent many hours above the Earth in flight. And I know "above" is not quite right. As much as I am enamored of rock, I realize that Earth is more than crust, mantle, core. As you observed in your first letter, Earth also encompasses its life-generating, planetary pond. If we want to think about the Earth as an object, then it must also include a layered atmosphere full of currents, gases, organisms, auroras, and airplanes. That extends the size of the thing we are talking about by a good 10,000 km, doesn't it? And it is not as if the demarcation between the exosphere and outer space is firm. When viewed from space the object Earth seems integral, but its borders are a mess. Its scale challenges. So does the adaptability of the life that thrives on its surface. I was just reading about tardigrades, otherworldy little microscopic "water bears," being able to survive in space. No doubt we've already sent some there on our machines, along with all kinds of single-celled organisms.

My favorite moment of airplane flight is takeoff, just as the nose lifts and it seems you are escaping something. I like to look out the window and watch the ascent through cloud cover, to peer at the expanded horizon. The number of miles that may be traversed along the curve of the Earth's surface during fifteen hours in a sealed metal tube amazes me, even as I realize that compared to a moon-bound astronaut or the New Horizons spacecraft approaching Pluto I have barely moved. And speaking of metal and distances, while we have been dispatching these letters this autumn your proposal to send a robotic spacecraft to 16 Psyche has been designated a semifinalist for NASA Discovery funding. Congratulations! What an alluring name that asteroid bears. As I understand it, Psyche may be the metallic remnant of a protoplanet's core, and might therefore offer a glimpse of the Earth's distant past. I would love to hear more from you about why this mission matters, and what (speaking of scales of time) is gained by knowing Psyche. How can probing an inhospitable space object open new knowledge about Earth? Does travel beyond Earth's atmosphere always aspire to be travel back in Earth's time?

You asked me about the challenge of scale in human history: do a thousand years offer the same cognitive stumbling block as 100,000 kilometers? That geographical span amounts to about two and half times the circumference of the Earth. Even to someone who has undertaken a great deal of travel, that distance seems *very* far indeed. But not exactly incomprehensible. The distance from Washington, DC to Auckland is 27,744 kilometers, after all, so if I flew to New Zealand and back three times and then once more to remain (about 3.6 trips, so maybe I would finally visit Easter Island en route), then I would be nearing the 100,000 kilometer mark. And completely exhausted, and hungry for something better than bland food heated in aluminum foil. But I would not be completing a journey that rivals anything like what the crews of Air New Zealand jets frequently and casually undertake. So, I can understand vast extension by reducing it to my own experiential units, just as you observed about measuring horses with hands and fields with strides. A meter is also a rather human span, about an arm's length. I wonder, does a kilometer or 100,000 kilometers then propel us out of our human frame? Yes, I know, they are based on units of ten, and that has more to do with the number of fingers we possess than anything mathematically natural—but I still wonder, do these capacious units give us a way of conceiving the Earth that doesn't diminish it into conceptual plaything, a Blue Marble? Or must we reduce vast extensions of time and space to small human measures to speak of them comfortably? If so, the Earth is surely outside our grasp, despite those photographs taken from the Space Station, satellites, the Moon. So much trouble arises when we try to position ourselves above or outside the Earth. We are always a part of this object we are trying to think about as if we were not.

As I did my calculations of planetary circumferences and plane to trips to New Zealand, trying to try to wrap my mind around 100,000 kilometers, I also wondered about the centuries before jet flight, and how vast distances might have posed a different kind of challenge (or at least a challenge to be differently reckoned with). To settle the islands of New Zealand and Hawaii, the ancestors of modern-day Polynesian peoples like the Maori journeyed more than 7,000 kilometers. They crossed these long expanses of Earth as leagues of open ocean, afloat in fleets of canoes. Such distances surely did not feel to these sailors as they do to an airborne traveler. More watery, more uncertain, riding the peaks and troughs of waves, always in the midst of things, probably not pretending it is possible to be above it all (but who knows? The stars are above it all). When you set your course for lands that that may not exist—as the first among these adventurers surely did—I am guessing that you would not conceptualize time and journey in the same way as a pilgrim voyaging to a known destination, especially a pilgrim with a good map and flight crew. Planets, the Sun, and constellations anchored their canoes to the Earth and let these sailors know about how far and in what direction they had voyaged. I speculate that it was easier for them to track their distances in elapsed time rather than through place-bound measures. But then again, we are forever underestimating the navigational sophistication of peoples who do not have our compasses, astrolabes, and GPS devices.

There's more to say about Earth's vast spatial expanses and what peoples in other centuries made of such conceptual challenges. In a future letter I promise to write to you about the fourteenth-century man who *almost* circumambulated the globe, but turned back just before he reached home. But I want to return in this one to the question you raised about temporal scale and comprehensibility. In researching my recent book on stone I was struck by how the origin of modern geology is always narrated as a rupture in history, the advent of a wholly new knowledge that changes everything preceding—almost like a detonation, an extinction wave, an asteroid slamming Earth and taking with it all the dinosaurs (or at least superseded knowledge). At the discovery of stone's profound age, time elongates into dizzying stretches and becomes alien, a world without us, an Earth on which we barely register. Planetary history had previously seemed rather shallow, almost wholly congruent with human history. Geology then declares that we cannot well apprehend Earth's temporal immensity. Arriving in crisis, an existential threat to human importance, the science of geology obliterates earlier, more mythic ways of knowing the world, at least for those who take its findings seriously. Geological time, we are told, blasts away the foundations of human-centered reckonings of history. Of course, this realization had to happen multiple times. The most famous is James Hutton (1726–97) realizing that the Earth is both exceedingly ancient and caught perpetually in a process of self-creation. In rock formations like Edinburgh's Salisbury Crags, Hutton read a story of stone's energy and constant movement, one in which horizontal layers of sedimentation bend, ripple, melt, interpenetrate. The story these rocks tell makes even the cataclysm of the biblical Great Flood seem the briefest of events, a small narrative to which the geological record might remain indifferent. Hutton was not the first to argue for a very long history for Earth. Theophrastus, Aristotle, Pliny, Al-Biruni, Ibn Sina, Shen Kuo, Jean Buridan, and many others are part of that complicated story of thinking the planet's immense duration. Yet Hutton is usually credited for unleashing upon modernity the challenge of understanding Deep Time, a cosmic timeline in which humans minimally feature.

A "time revolution" unfolds in the nineteenth century, driven by geological discoveries like Hutton's. Short histories of the planet deriving from Genesis find themselves undermined, discarded. In the seventeenth century, Bishop James Ussher infamously calculated an origin date of 4004 BCE for the Earth based on his reading the biblical narrative. This strangely precise number has little historical precedent, though, especially because literalism is in general a fairly recent mode of interpreting scripture. Believing the Bible to offer exact history, a storehouse of fact, contrasts greatly with Jewish and early Christian methods of interpreting stories for meaning and symbol. A day of creation might designate eons, especially since the sun and moon do not exist until day four (and who knows what a day might mean to a deity existing before and outside of time?) Calculations of the Earth's age were therefore not all that important—and varied widely. In the Middle Ages, for example, calculations of the span between day one of creation and the present ranged from about seven thousand years to a very precise 3,852. The numbers shifted because the Bible does not offer a straightforward chronology. Yet all these reckonings offer thousands of years, not billions, a measure that strikes us now as impoverished. Geology insists upon the eon, era, period, and epoch. The past revealed in stone's archive is so extensive that a new era of human relation to history is inaugurated, one in which temporal extension is so profound that it makes the intellect reel. Gone are the days of the accessible, comfortably human century and millennium, so much easier to grasp.

But is a thousand years really so unchallenging? A millennium ago Cnut of Denmark was leading a victorious invasion of England. It's easy to forget that in the aftermath of his campaign the kingdom had three Danish monarchs sit upon its throne and was for a while annexed to a North Sea empire. Though profoundly important in his own time, King Cnut is now mostly forgotten to history. Some people recognize his name from a famous story about how even though he commanded the ocean to stop rising, the advancing tide soaked his feet all the same. The tale is almost always told with the wrong moral, as if it illustrates the vanity of a ruler who thought the elements should obey his earthly authority. As the twelfth-century historian Henry of Huntingdon makes clear, Cnut held court by the sea as an act of public humility.[3] He wanted to emphasize to his nobles the limitations of human power, even that of a monarch's, over the world in which we dwell. He wanted to teach them that the Earth exceeds us. I wonder if a millennium is so easy to hold and know as we think it is. A thousand years can make most human things vanish, swallow them in a tide ("tide" is the Old English word for "time"). No one names their children Cnut anymore.

Methuselah is the longest-lived person in the Bible. He endured 976 years, just short of the millennium. Had he lived much longer he might have been admitted to the ark of his grandson Noah (though knowing Noah, Methuselah may also have been left outside). Exceeding most human lifespans, a century is challenging enough to wrap your mind around. Most people blur together their own past, especially when they reach—*ahem*—a certain age. How much has the world changed since 1915? Since 1985? The thousand years to which Methuselah almost survives is only relatively easier to grasp than geology's million- and billion-year parceling of time. That's why, I think, it is almost impossible not to insert a familiar observer into any scene from a time whose distance from us leaves us flustered: the inveterate mistake of adding cavemen to pictures of dinosaurs when thinking back through history, or our difficulty in imagining a future for the Earth in which we are not abidingly present. That's not to say we are not capable of all such distancing. Alan Weisman's *The Earth Without Us*, for example, envisions the slow collapse of our infrastructure once we are no longer terrestrial residents, the creation of fossils and geological seams long after humans no longer thrive.[4] Then again, even in this imagining we are in a way still present, since Weisman narrates such far futures as if witnessed by an observing human.

In writing of King Cnut I mentioned Henry of Huntingdon, an English historian who lived in the twelfth century. At the end of his book Henry composed a letter to readers living millennia after his death, asking them to contemplate how quickly historical memory recedes. In thinking back a thousand years, Henry was struck by the scantiness of the historical record, preserving as it does little trace of most human and animal lives. Henry asks readers a millennium after his death—*us!*—to contemplate the obscurity into which his times have surely fallen. He urges us to contemplate the fate of our own biographies, the destiny of our works. Henry next addresses readers two and then three millennia beyond the finishing of his book, should any humans still dwell on Earth. Of these far distant audiences he asks a favor: to remember and to pray for him, and to ponder all those who have been swallowed by time's silencing immensity. Henry's request of this far future seems humble, even sweet. And yet as he addresses his imagined audience of 4150 CE, he betrays a quiet confidence that if humans should still inhabit the Earth three thousand years after his own demise, they will be reading his book and enjoying his lively Latin prose.

Well, maybe that is Henry's leap of faith. He has no certainty that the Earth will still be populated three millennia after he is dead. The apocalypse may well have come and gone. Yet he nonetheless mails a letter to those future readers, trusting it will arrive somewhere, trusting his words and his imagination can provoke a community across unlivable spans of time. I do not have any such hope that these letters we are sending back and forth will endure more than a few years, Lindy, and that seems OK to me. Unlike Henry of Huntingdon I believe our endeavor to be more modest. We are just trying to describe what is at stake in contemplating the Earth as object, and in doing so attempting to convey immensities of scale (of time, of distance, of temperature) no human can really know. Easy stuff! So tell me, in thinking about your 16 Psyche mission, does that massive asteroid seem close to you, or impossibly far? What's at stake when we talk about scales of distance? Can we ever know the Earth in its enormity, let alone comprehend the trek to a "nearby" cosmic object?

—Jeffrey

PS This letter took me so long to compose that my birthday has come and gone. I'm now one year older than you. I can tell you that I have seen your likely future and it contains community, some good cheer, and inevitable propulsion through time. And also, cake.

Dear Jeffrey,

Tonight I am sitting in our perfectly quiet house with the Arizona night sky visible out the windows. There is prosecco by my side and I have had some jelly beans, too. I've spent the day at the Mayo Clinic, as I do every three months now, getting blood drawn and having long strange MRI scans where I have to breathe in and breathe out and hold my breath over and over for almost an hour. I am nearing the one-year anniversary of my last chemotherapy. I expect the result of the tests, which I will receive tomorrow, to show that I am cancer-free. And I expect that same result for the rest of my life, because I am in a very fortunate small group of women who had their ovarian cancer detected very early, by chance, by luck.

And as common and frankly clichéd as my story is, it has affected my sense of time. When my doctor first told me I had cancer, I quickly checked myself to see how I was reacting. Yes, I'm always over-thinking, often meta-thinking, the very walking definition of irony. In fact in that moment I discovered that I was not panicking and I was not experiencing the cold rising tide of regrets. I discovered that I feel that I have been fully engaged with my life all along, for better and for worse, for richer and poorer, and now in sickness. There were plenty of moments I'd like to do over and plenty of hurts I have doled out that I would like to take back, but I did not have the nauseating feeling that I had wasted any time. So, on to surgery and the brutal almost-death that is chemo "therapy," with a clear heart. At least, clear-ish.

It's now, in the aftermath, that time has changed for me. Now that I am feeling pretty good again and am back in high gear with my job, which I love. Time with my family has been constant and intensely rewarding and very warm all along through this year and a half, but my work was triaged in the most heartless way when I had so little energy to think. Now I'm back on my game! And all kinds of fantastic things are happening. Working with my colleagues in the School of Earth and Space Exploration is actually a daily pleasure, something rare in academia. My son and husband and I have a brand-new startup company called Beagle Learning. And I am working on this book with you, and I'm in a leadership-training course, and then there is the NASA mission proposal to visit Psyche. Suddenly time feels short and precious.

What if the scans show a return of the cancer? Then the research says my odds are very poor. Then I will not get to spend enough time with my family. I'll miss some decades we'd have been together. And I won't be able to work on these wonderful big projects and try to change the world. What if I only have a couple of years, instead of three or even four more decades? That would be a disappointment indeed.

I'm not, at the moment at least, afraid of dying, perhaps because I don't at all think I need to face the bad news I am hypothesizing. But it demonstrates the severe limitations we have with both time and space. When threatened, my time threshold shrinks back to just this year, or just this week, or when I was in a lot of pain: just get through this moment. And in those times the only space I inhabited was my body, and no further. With these limitations, it is astonishing we can ever begin to comprehend the size of the solar system or its age. In any case, it helps to have a healthy body to free the mind do it! Thank goodness that is the state I am in now.

Time. You are right that Hutton sparked a lot of the understanding we have today (and Hutton has that poetic quote, "We find no vestige of a beginning,—no prospect of an end."). But I lay more of the glory upon Charles Lyell. Lyell set out to write books that would convince the populace of his findings, something we do all too seldom these days. And one he helped convince was Charles Darwin.[5] Lyell solved Darwin's time problem: How could living things change so fast? Answer: they are not changing fast. They are changing slowly, over far greater spans of time than we had thought.

Darwin wrote, in *The Origin of Species*, "It is hardly possible for me to recall to the reader who is not a practical geologist, the facts leading the mind feebly to comprehend the lapse of time. He who can read Sir Charles Lyell's grand work on the Principles of Geology, which the future historian will recognise as having produced a revolution in natural science, and yet does not admit how vast have been the past periods of time, may at once close this volume."

Prescient he may have been, with the numbers of people on Earth today who choose to close that volume. How, Jeffrey, do we convince people that they should not hold opinions based on emotion, but on information?

Darwin also dedicated the second edition of *The Origin of Species* to Lyell, writing, "To Charles Lyell, Esq., F.R.S.

The second edition is dedicated with grateful pleasure, as an acknowledgement that the chief part of whatever scientific merit this journal and the other works of the author may possess, has been derived from studying the well-known and admirable PRINCIPLES OF GEOLOGY."

Lyell had studied with a careful eye and a thinking mind many geological structures, including Niagara Falls and eroded portions of the Alps. He calculated that Niagara Falls had been eroding back for 35,000 years. He posited that the Alps had experienced entirely different climates, from glacial to temperate. These evidences are all around us. Darwin took Lyell's first volume with him on the *Beagle*, and he awaited the second volume eagerly. His sister shipped it to him when it was published and he received it while in South America.

How long, though, is long? Archbishop Ussher still has his adherents, though as early as 1748 Benoit de Maillet suggested the Earth was more than 2 billion years old. Over a century later, in the 1860s, Lord Kelvin tried repeatedly to use heat loss to calculate age. Starting from today's temperatures, he calculated back to a posited blazing hot origin, using the assumption that heat is conducted through the rock to the surface. Lord Kelvin produced results between 24 and 400 million years for the age of the Earth.

Lord Kelvin was wrong first because much of the heat in the Earth is not conducted to the surface (moved by atomic vibration through a solid, unmoving material, in the way that heat comes through a coffee cup) but instead through convection. In convection, the hot material itself moves and delivers its heat to the surface. The mantle of the Earth, under the brittle tectonic plates, is very slowly convecting, like a pot of oatmeal set to one-billionth speed. The mantle is solid rock, but it moves very slowly through deforming its crystals. Convection moves heat far more quickly than conduction does.

Secondly, Lord Kelvin was wrong because he did not know about radioactivity. No one did; it was not discovered until 1895, and its importance in constantly heating the Earth's interior is still being actively studied now. The final big leap in determining the age of the Earth was made in the 1950s when scientists discovered how to use radiodecay of elements trapped within minerals to determine when the minerals were formed. In 1953, F. G. Houtermans and Clair Patterson set the age of the Earth at 4.5 billion years. With the increased nuance that further study has brought, we say that the first solids formed in the disk around the infant Sun 4.567 billion years ago, and the Earth accreted to its present size over about the next 100 million years.

I love this story for its humanness. We think and think about whatever is in front of us, be it Niagara Falls, or a piece of uraninite, or a Bible. And the best answers are made when we bring all the information together.

This letter has gone on far too long. It's been a pleasure for me this late night to travel from the fraught subject of my health with its fearsome imprisonment in a tiny space (my body) and consciousness for too brief a time (my life), to the beauty that can be achieved through thought. Can I truly understand 4.567 billion any better now, for two decades of handling it regularly, than I could before? No, and as my husband James Tanton says, familiarity is not the same as understanding.

You wisely asked, "Does travel beyond our exosphere always aspire to be travel back in Earth's time?" but I will hold back my thoughts on that, and on Psyche, until another letter.

I hope this letter finds you well and close to your family. Send them all my very best. And writing to you from my fifty-first year to your fifty-second (am I right?), I am happy to report that there is community, good cheer, and, thank goodness, cake, in this year as well.

—Lindy

Dear Lindy,

Congratulations on your clean bill of health! When we saw the good news you posted on Facebook about being cancer-free, we poured some wine and toasted your recovery. We are so happy to be on this Earth with you.

Your meditation on mortality, being an embodied and imperiled observer, relations of love and regret, how time expands and contracts depending on context or the point from which its unfolding is observed . . . well, your last letter gave me much to think about, and I'm grateful. You composed in the company of prosecco and jelly beans (I am imagining you placed a green one in your glass of bubbly to serve as an olive). A gray November morning is just commencing here in DC, so I'm at my laptop with strong coffee and a totem by my side: a small chunk of rock that you gave me the first time you visited the Cohen household, back when we did not live all that far from each other. You and James came for dinner bearing a bottle of prosecco and a vibrant slice of stone resembling a piece of birthday cake. Layers of brown alternate with white, and a stripe of radiant "frosting" glazes its top. When I hold the rock in my hand (it is palm size, one of those stones that yearns to be grasped) I think about time: how long was this rock in the making? I think about temperature and crushing force: what pressures and what heat were required to meld its layered particles into such density? I ponder vast extensions of space (how far did those sediments travel before becoming this slice of lithic cake?), beauty (the rock shimmers and I have to hold it), and swirling water (surely the strata of the stone are an archive of its intimacy with that element). And birthdays, since it really does look like cake. All the themes of these letters seem condensed in this little fragment of the Earth that most days sits on our windowsill and reminds me of friendship.

But sometimes, like this morning, I can't help palming the stone. It's an action I undertake without forethought. To my surprise I am in the motion and then the object is suddenly resting in my hand. I've repeated a similar gesture throughout my life, most often with wave-smoothed pebbles that look like eggs or little earths, most often when beachcombing in Maine (my father's family is from that state, and I often think my love of rocks and crashing water are the impress of its wild shores). But I'll absentmindedly scoop up most any stone of the proper size, contours, and allure. The Cohen house is filled with petric souvenirs. Not all of them are mine. From earliest childhood both our son and daughter have been lithophiles, filling their pockets with little chips of the Earth on our neighborhood walks. To prevent our home from feeling like a quarry we used to tell our children to empty their pockets on the front steps before entering. That rule ensured small cairns would always be a part of the landscape of our front yard! I miss them, now.

I suppose I shouldn't over romanticize such things, but it seems to me there is something in stone that wants to be touched, some offer of mineral friendship. Or at least an invitation to think about hard and enduring things, about inhuman durations of space and time. I like the geologist Jan Zalasiewicz's[6] observation that a "pebble holds strange worlds within it." I'm a big fan of his *Planet in a Pebble*. The book moves with dizzying rapidity from a disk of white-striped gray flint he finds along the beach in Wales to the Big Bang and the distant future. Zalasiewicz reveals how the Earth's expansive story is discernible within its small, ubiquitous fragments. The narrative he unfolds from the stone in his palm is an adventure story in which cosmic origins, the fossils of ancient organisms (billions of corpses from the Silurian sea), and the plowing of ship-like continents along an ocean of molten rock all figure. His pebble comes from the now-lost microcontinent of Avalonia, "one on which—much later—King Arthur would reign, and Shakespeare would write sonnets, and a revolution that would spread factory chimneys and iron foundries across the world" (33). I love that *Avalonia* is the name we have given to this ancient expanse that in time became Wales and England—but also scattered into stony terrain found throughout Western Europe, Africa, and North America. Through a name like Avalonia we attempt to make an unreachable past knowable, adapting a land lost in time to our beloved myths, as if Morgan le Fay might have once taken Arthur there. Likewise we label the Earth's first geological epoch the Hadean, as if we could know that billion and a half years of intolerable heat and poisonous atmosphere from stories long told of unlivable underground domains, of lands beyond human experience but not beyond our imagination and dreams.

James is right to say that familiarity is not the same as understanding. The Hadean is not Hades, a place the Greeks imagined might be visited and survived by a resolute Ulysses. I thought about James's observation when I grasped the birthday cake stone before writing this. I am familiar with it, of course, since I see it every day, but I can't claim to understand its formation or its geological time (even as I try). Or maybe I keep picking it up because I realize that even after two years I am still not so familiar with the stone that I no longer see it. The rock hasn't gone invisible like so many of our other tchotchkes (it sits on a windowsill with a tiki shot glass, a candle holder, and two goblets that our daughter made at pottery camp—but I had to look up to remember that). I wonder if the birthday cake stone's beauty is not a guard against familiarity. The rock shimmers and I pick it up. An allurement to desire. I wonder if what is at play in my household palming of the stone before writing you a letter—or in Jan Zalasiewicz picking up a pebble along a Welsh beach and discerning within it the long story of the Earth—isn't the cognitive gravity of beautiful things, their ability to pull an observer away from familiarity and toward encounter with the unknown, the hard-to-grasp. Beauty would then entail recognition of our ignorance, a turning toward the unfamiliar, an impulse to *grasp* (in the double sense of holding in hand and holding in mind). That desire might never be fulfilled. Maybe that makes it become all the more intense. *Aesthetics*, the study of beauty, comes from the Greek word for perceptible or palpable things. It's about bodies and touching as much as thinking and feeling. Beauty is a prod to desire, and desire can be an invitation to knowing, to grasping. Under the right conditions that pull can yank us from the orbit of the merely human so that we attempt to understand scales of distance and time that exceed us as ephemeral, embodied creatures. A pebble can be sublime.

So can the Earth in its entirety—and that radiance is both an invitation (the Blue Marble as a planet to cherish rather than exploit) and a problem (the Blue Marble as toy for human mastery and use). Humans have long been striving to imagine what the Earth might look like if viewed from above, if beheld through the eye of a god. Medieval theologians, encyclopedia writers, and historians postulated that the chaos of the turbulent globe would resolve into gorgeous order, if only glimpsed from a celestial perspective. In the fourteenth century Geoffrey Chaucer concluded his narrative poem *Troilus and Criseyde* with the death of its hero, a Trojan warrior filled with despair after his beloved abandons him for another man.[7] Troilus dies, despondent, in combat. As his soul rises through the Earth's atmosphere, "leaving every element" behind, he glimpses the wandering stars. His ears fill with a planetary melody, and he turns back toward the receding Earth, toward all that he has until this moment known. He contemplates the surprising smallness of his former home, a dwindling expanse of land and water: "This litel spot of erthe, that with the see/Embraced is" ("this little spot of earth, that is with the sea embraced," *Troilus and Criseyde* V.1814–17).

As he rises beyond his diminished home, a world that when he lived there seemed so vast, Troilus realizes the minute scale of his own life. Everything that seemed important, filled with passion, so grand, diminishes into a spot of earth and sea. Troilus laughs.

Viewing the world from its imagined exterior, watching immensity recede into small prospect, deflates human self-importance. The moral of such stories is usually that humans do not matter all that much. Our lives are too brief, our powers too limited. The Earth's enormity is only relative, and transcendable. I wonder though if that isn't just another tale we tell ourselves to excuse our devastation of the Earth. We invoke cosmic scale so we can go on with business as usual, unperturbed. Yet it's worth noting that Jan Zalasiewicz, the palaeobiologist who composed *Planet in a Pebble* and who recognized in a stone held in his palm an invitation to imagine the origin and fate of the Earth, is also the convener of the Working Group on the Anthropocene. This committee will recommend to the International Commission on Stratigraphy whether the Anthropocene—the Age of the Human—is a geologically recognizable era. Various starting dates have been proposed, with most clustering around the time of the Industrial Revolution or the Nuclear Age. At stake in embracing the term Anthropocene though is the recognition (or not) that humans have so profoundly altered the Earth that their activities and dubious achievements are now readable in the geological record. If the Anthropocene is officially accepted as a discernable-in-stone designation, it will name the shortest of geologically significant intervals. I'd be curious to hear your thoughts about the Anthropocene, Lindy, and maybe in a future letter I will write about who and what gets left out by its "Anthro-."

But I want to close this one with the story I promised last time. It's from a book of mysterious origin typically called *Mandeville's Travels* (most medieval books did not have titles, so what we call them tends to be a modern invention).[8] The book appeared suddenly in the middle of the fourteenth century and has been immensely popular ever since. Christopher Columbus even carried a copy on his voyages. John Mandeville tells us that he is an English knight born in St. Albans, a town near London known for its monastery and map making. His story begins with a pilgrimage to Jerusalem, which he identifies as the center of the world. He seems to be thinking here of what are called T-O maps, where the Earth is depicted as if it were a sea-clasped circle (the O) with three bodies of water (the T) dividing its landmass into Europe, Asia, and Africa. T-O maps are conceptual rather than practical, but Mandeville begins his journey as if moving across a globular version of one in which every approach to the Holy Land is uphill. Interestingly, though, once he completes his pilgrimage John Mandeville keeps on trekking.

A travelogue narrated with vim, the *Mandeville's Travels* enables us to wander the world in Sir John's excellent company. He's quite a companion: knowledgeable in good wine, balm, and diamonds; happy to advise us where translators will be needed and what alternative routes we might take if we wish to follow him to the Holy Land; eager to recount stories about the marvelous denizens of Asia, India, Africa, and many unknown islands. Few things anger or perturb him. Rituals, beliefs, travel of all kinds, diverse peoples and strange objects utterly fascinate him. Rather than be repulsed by anthropophagy or sky burial, for example, Sir John explicates the cultural logic behind such practices (honoring the dead perhaps, so that they will not lie in the cold earth or be eaten by worms). No rite or custom seems too outré. Even the nudist communist polygamous cannibals of the island of Lamoria behave within a sensical system of belief, with each of their choices making rational sense. Only the landless Jews earn Mandeville's ire (cosmopolitanism always has its limits).

Next to the Bible, *Mandeville's Travels* was the most popular book to have circulated during the Western Middle Ages. It exists in numerous languages and manifold versions. The world that John Mandeville opened was alluring. He traveled farther than most could imagine. The reports he brought back of idol worship, hermaphrodites, Amazons, assassins, and holy shrines were narrated from his own experience: he serves as a mercenary for a Muslim sultan who provides him an all-access pass for his extensive dominions (and offers him a wife as well), and learns the ways of the great Khan's court in China by residing there for a while, before journeying to Tibet. Sir John tells us that his desire to wander the world's immensity was triggered early in life, when he heard the story of a man who nearly circumambulated the globe. This traveler passed through India and the vast lands beyond, ever pushing forward, until he came to a strange island where the native language sounded almost like familiar English. He could find no further conveyance, though, and so reversed his voyage—a very long way home. Later, having returned to England but too restless to stay long, the traveler took ship and was blown off course in a storm. The vessel arrived at the very island where he had long ago turned back. It turned out to be not all that far from home.

I love three things about this tale. First, its arduous journeying incites the young Mandeville to his own peregrinations: John Mandeville, like this unnamed traveler, enjoys travel more than destination. Second, although Mandeville emphasizes the roundness and the immensity of the world, he can tell no story of anyone having successfully completed a voyage that traverses the Earth's entire circumference. You cannot arrive home by pressing only forward. Mandeville's world is not fully to be grasped, not fully to be known. There will always be something that prevents or escapes encompassing—and detours well worth the wandering. Last, there never really was a Sir John Mandeville. His travels are something of a literary hoax, composed by an unknown author and based on texts written by many other people—some of them real travelers, some not. The book is an act of reverent plagiarism in an age without copyright. The traveler who nearly circumambulates the Earth is no doubt an invented person, yet another fiction passed off as enticing fact, as much a work of the imagination as Sir John Mandeville himself surely is. And yet the fact that the *Mandeville's Travels* is a conceptual journey that strives for tolerance and disorients everyday perspectives is what makes the narrative so powerful. I don't need a real John Mandeville to have existed in order to know that his refusal to reduce the world to a small and knowable "little spot" matters.

Time has become space. Or, rather, it is difficult to separate those scales. I've worked on this letter over the course of a long day. Good coffee became mint tea, and now it is early evening. I am thinking prosecco next, though probably not with jelly beans.

—Jeffrey

5 RADIANCE (EARTH'S BEAUTY)

December 1, 2015
Instant messages exchanged as we sit in the same room, Lindy Elkins-Tanton's office at the ASU School of Earth and Space Exploration. Tempe, Arizona.

> Jeffrey, I'm so pleased to be reprising today our conversation from years ago. I wonder how our thoughts will have changed? And I have to start as I did back then, asking you, what are the big questions you are hoping to answer with your work?

I knew you were going to ask me that! First, let me say that it is good to be sharing space with you as we chat by typing to each other across this comfortable room. I am sitting on a couch in your office at Arizona State University, and you're at a stately standing desk. It is good as well to have a chance to contemplate the stakes of this book, something we've been working on together for a long time. So. With *Earth* my big question—and maybe our Big Question—is: can a collaboration that brings together a scholar trained in the natural sciences and one trained in a precise corner of the humanities somehow convey to an interested public what the Earth (as an object) might be? What are the risks and rewards of thinking about this thing so inhumanly vast it escapes every scale of our comprehension?

This is a new kind of writing for me, like a slow conversation. Faster than letters, slower than speech. I think we are on to something. Risks and rewards. The notion of there being risks is interesting to me. Were I pre-tenure, my chair might say that I should focus on research that addresses a strictly scientific, testable question and will result in a high-impact journal publication in my field. Our work here would not fulfill that goal. But now, as befits the academic system, as tenured director of a school I am free to pursue these meta, cross-disciplinary questions. What other risks are there? Might we run into that strict definition of the sublime, that we are overcome with the immensity and incomprehensibility of our universe? That we are crushed under the weight of our human meaninglessness? Perhaps I am too small to be crushed, like that industrious ant that crawls unharmed from under a boot.

Neither of our disciplines necessarily rewards public writing of this kind, but a book about Earth has to aim for a wider audience than a peer-reviewed journal enables! Earth is an object of such immensity that it beckons us to think outside of the comfortable orbits of career path and discipline. The risks must also include that we would talk at each other (your Earth might be a planet with a history so long that humans simply do not figure much; mine might be an Earth that is too much a home for humans and does not adequately engage with the longevity of stones, water, or atmospheric gases). And maybe the sublime is a risk in another way: that if it opens us to human insignificance in the cosmos (that question of scale again!) then we forget that feeling overwhelmed is a recurrent and maybe even transhistorical emotion. Emotion is a spur to cognition. We write this together because we have been intellectually and affectively spurred by the difficulty we have in grasping this object we are writing about together. So in order not to be crushed by the weight of the Earth (we can't presume to be Atlas) we are mapping multiple routes into comprehending this planet as an object and attempting to convey why such comprehension matters. So, what are the rewards? Given that our question is not testable or solvable, how will we know if we are succeeding?

Oh my god! Is there such a thing as a nontranshistorical emotion? Are there emotions that we have now that earlier people never had, or vice versa? The risk of talking across each other was a concern to me when we first began our collaboration, years ago, and I realized that the humanists at the conference I attended were aiming to have as many layered meanings and evocations in their sentences as possible (is that "polysemy"?) while I, speaking as a scientist, was trying to have exactly one meaning for each sentence, and to have each person in the room grasp that meaning. I have found, though, that we do converge and speak to each other, though sometimes the convergence requires more than one sentence. Perhaps this is a lesson for all people trying to communicate. And so, rewards. I feel the reward as a burst of excitement as I find a new understanding of a familiar topic. It's that addictive experience of discovery and the writing is the exploration that leads to the discovery. There is a reward in allowing myself to stray from the strict world of data and interpretation into my human experience of science, which includes the beautiful and the personal.

So let me ask you about the beautiful first, and the personal after that. First, the beautiful: I understand that data is convincing when it is testable, repeatable, represents a sufficient sample size, and so forth. I understand as well how data is communicated in, say, a peer-reviewed journal about planetary geology through a shared language (partly mathematical, partly English with more precise definitions adhering to key terms—e.g., last night at dinner we talked about the world of difference between what a planetary scientist means when she speaks about dust and ash and what an everyday speaker understands by those terms). It takes a great deal of training to be able to comprehend data and its interpretation and then to decide if such data is compelling. Yet last night you also showed me the cover to your finalist proposal for a mission to (16) Psyche, a bright rendition of two planetesimals smashing together to form the metal asteroid, full of fiery orange and hot white and gray rocks melting into red. Even the title of the proposal is evocative: *Psyche: Journey to a Metal World.* When I read that I wanted to say, sign me up! So I would like to ask, what is the function of beauty in scientific persuasion? We've spoken in our letters about luminous images of the Earth from space and the work they might do to convince people to think or feel differently about the planet. As a planetary geophysicist, what is the function of beauty in your work?

Beauty. OK, it's taken us 15,000 words to lay this question out starkly, and here it is. A little bit tongue-in-cheek, I would say that pre-tenure scientific faculty do not speak about beauty very much (too personal! irrelevant! soft!) but the conversation thickens over time. In a broad sense beauty is the reason we all do what we do, why we choose the field we choose, and what gives us the determination to keep working in the face of discouragements. We do not choose to study the minerals in lava flows because it was an obvious career alternative to the law, we do it because lava is beautiful and the notion of interrogating the untouchable heart of the planet through incandescent molten eruptions appeals to our unconscious understanding of human relations and our place in them. We do not choose to study subatomic particles because of the job prospects, but because understanding the workings of atoms gives us a joyous grasp of the stability and architecture of the universe, and that helps us feel included and enduring. These motivations are all aspects of beauty.

Beauty is what propels desire, and joy, and an impulse to know. I can't disagree. In a letter to you earlier in this book I described how beauty invites the hand to grasp before the mind apprehends the motion of palming a smooth pebble. I wrote a book on stone because its beauty won't leave me alone. I also think a great deal about the relation of medieval people to the planet they inhabited. On the one hand the cosmological diagrams they drew are wrong: the Earth is at the center! The moon and the sun are orbiting around the planet at nearly the same distance! There are not enough planets! There's a god or some angels placed in Outer Space! And yet those concentric rings revolving around each other, drawn with precision and often vibrant in their colors, make we want to know more about this worldview, and maybe even recover from it something that is not error, something that might be a spur to telling better stories about our Earth now. And it is interesting that your answer about science and beauty is full of Earth fragments (lava, the molten planetary heart, subatomic particles) and personal fragments (not everyone gets to choose between law and physics, for example—and not everyone is drawn to rocks or to inhuman forces). We write our own experience best. We are embodied and we compose from our lived perspective. But I am also wondering how we widen that personal point of view to embrace objects on a scale that exceed us and yet guard ourselves from losing the words to narrate or the desire to connect. I wonder how we trigger communal widenings of perspective in the hope of a more just or at least more sustainable world. Let me ask this specifically of Earth: do beautiful images of the planet impede or enable?

I suddenly see the old maps of the cosmos with Earth as the center a more true representation of our experience, in a way I have not before. I do think that the same unconscious constructs that can impel us to therapy also drive our passions and what we find beautiful. Devoting one's life to the pursuit of knowledge and education, as an academic often does, or to a singular art, or any intense, large, driven pursuit, is a pure reflection of the strength of the unconscious construct. So suddenly I see that the Earth-centered cosmos is not a simple scientific misunderstanding but a true expression of our lives inside our heads, of our necessarily individual, separate experiences. OK. So, to your question, do beautiful images impede or enable widening of communal determination to produce a more just and sustainable world (in my paraphrase)? I suspect they may impede, unless connected to a far broader dialogue. What does a beautiful image do but connect to and thus reinforce our existing interior constructs and beliefs? How to share our sense of the *meaning* of the beauty, the message of the beauty. How to share, with those who don't immediately apprehend this themselves, that our beautiful little world can be better. We can so easily ruin this world for ourselves (though the world and other life will persist, no worries!). We can also so easily make improvements that lead to less suffering. I am not sure an image of the world speaks those words to most people.

Sometimes, in fact, the desire to see the Earth as if you are not living upon it can be a way of escaping those tough questions: pretending you can inhabit a disembodied perspective, pretending you are not part of the object you behold. Can we really escape the planet on which we dwell? Can we know Earth more factually, more dispassionately, when we look back from the Moon? This morning when you were on a teleconference and I was sitting in the visiting faculty office, thinking about what we might chat about today, I found myself re-reading Cicero's *Dream of Scipio*, in which a Roman general dreams that he is lifted into space to look back upon the dwindled Earth. The Milky Way shimmers around him and he can see that his beloved city of Rome has from this distance shrunk to insignificance. Planetary spheres revolve and from this perspective Earth appears as a banded globe: snowfields at its polar regions and burning desert along its middle. Two temperate zones line either side of the torrid middle section. These two thin and fragile strips bounded by extremes of heat and cold are the only parts of the planet that can be made a home. This story survived by an accumulation of accidents, and was a spur to all kinds of geographical and cosmological inquiry in later ages, as well as to some beautiful images that make Earth resemble Jupiter with its colored stripes. In the Middle Ages, for example, the text of *Dream* was extant only within an extensive fifth-century commentary upon it by another Roman, Macrobius. The narrative tries to imagine an impossible perspective, one in which the Eternal City is just a dot, nothing all that important. The vastness of both Earth and the cosmos are overwhelming to Scipio, who awakens full of Stoic resignation to leading a good life in a world that utterly exceeds him. But that seems a very personal choice, one Roman general's decision of what to do when overwhelmed by cosmic beauty. No community is conjured up to try to make the Earth more habitable or just (in fact Scipio goes to Africa two years later and utterly destroys Carthage, razing its buildings and sowing its fields with salt). I guess I am wondering—since we have been thinking about emotions that cross the centuries—if the awe we might feel when we look upon the Earth as a radiant sphere (and humans have been imagining that perspective for millennia) is too often a prod

only to personal revelations and selfish resolutions.
Nothing changes.

> Astonishing. We need to start an encyclopedia of
> the responses to the realization that the Earth is
> a tiny speck, not the center of the universe, and
> that natural disasters occur with an impersonal
> randomness, and that we are microscopically small.
> In James Secord's introduction to the Penguin
> edition of Charles Lyell's *Principles of Geology*,
> he talks about the effects of Lyell's work on the
> emotions of the time. George Eliot concludes *The
> Mill on the Floss* with a great flood that kills her
> characters, and she writes that its damage was soon
> overprinted with new trees and grass. This was
> part of a despondency that settled in the wake of
> *Principles of Geology* in those who felt that taking
> the times and events of the Earth out of God's hands
> and leaving them in a thoughtless cacophony of
> natural disasters similarly robbed our own lives of
> meaning. Why strive, why improve, when we can
> be wiped away by a random event, and so will all
> be wiped away by random events in the end? So,
> the very realizations that comfort me with the idea
> that our wrong acts are dissolved by time and their
> miniscule nature in the broader universe, and thus
> give me strength and courage to go forth and be able

to make mistakes and still persist, caused others to give up their striving.

And—as we just said to each other, because we are in the same room, chatting online but also sometimes speaking to each other in ways that don't get recorded here—that necessary encyclopedia requires an entry for the recurring human impulse to imagine the Earth as inscribed with art that you can only see if you are not an ordinary human living and dying on its surface. I'm thinking of those desert images called Nasca Lines, massive geoglyphs in Peru that depict animals, strange human forms, and geometric designs. They are only viewable as coherent images from far above . . . and so were created by imagining a perspective that could not be physically inhabited by their artists. Other such works exist all over the world, and continue to be constructed today (crop circles, anyone?) Yes they often get explained as visual offerings to gods or to aliens, but I think they are also transhistorical signals of the human desire to be above the planet we are bound to in life and in death. We send these messages to ourselves, and to those who come after us. It's about sharing a worldview, or an above-the-world view, POVs that are contagious and call for replication. We were just remarking on the perennial popularity of spaceflight films, which likewise leave

Earth behind (always with the obligatory shot of the planet receding into a tiny dot: Scipio's dream is alive and well these many centuries later). But these films also quietly arrive back on Earth again. Some earthbound landscape takes the place of the alien planet's surface for convenience of shooting— Iceland, for example (I think we have both been to the volcanic expanses around Hekla, where the *Alien* prequel *Prometheus* was filmed), or the deserts outside of Los Angeles, or even here in Arizona, where we are writing. From your office window I can see lines of scrubby brownish-green mountains that remind me of the geology of distant planets as imagined by Hollywood—mostly because so many films were shot in these dry, barren environs. But they are Earth all the same. This planet has such a hold on our imaginations that no matter where we travel we find ourselves back to some version of it again: its distant past, its anticipated future, or more likely a place currently existing but far enough away from the experience of most of the film's audience to seem like another world entirely. Don't you get tired of the SciFi convention of the Desert Planet, the Swamp Planet, the Ocean Planet—as if only one climate (Macrobius called them temperature zones) can exist on other worlds? Another inheritance of the *Dream of Scipio*, maybe. Do you think we are creating a kind of conceptual geoglyph through this book—and is that really the way to spur Earth thinking?

> As a quiet tired sigh at the end of this excellent writing session this morning, I will say here that we have reached the heart of the matter. Earth is an object and an icon not because we live here and it is our home, but because there is a universal, ageless, and completely bizarre drive in humans to see it from far, far above, as Scipio, Cicero, Macrobius, and the ancient peoples did, and in fact to leave it, to fly away into the universe as we do in our fiction and as we have begun to do in reality. Why would we have this wish? Perhaps for the simple reason that exploration is a deep human imperative. We are the apex predators, the owners, the graspers. And perhaps for an even more fundamental reason: Earth owns us with its gravity and its habitability, and these are shackles that are the most difficult of all to escape from, and so we wish to escape all the more fervently.

6 GRAVITY (EARTH'S PULL)

December 2, 2015
A Conversation. Tempe, Arizona.

Yesterday on the way home from campus we were talking about space travel's compelling and continuous presence in our imaginations. We were driving to your house in the dark—it had been a long day. The wrinkled hills loomed, spotted with cacti and strange outcroppings, and they seemed like an alien landscape. Again we found ourselves talking about why human beings throughout history have imagined themselves as being able to look back at a planet that had been their home, sometimes with longing, sometimes with affection, sometimes with horror at seeing something so important to them diminish. Why is it that we keep imagining this view? Why is Earth's gravity something we want to escape so badly?

I'm increasingly astonished as we talk about this, that humans imagine traveling through space at all. We sometimes think of imagining space travel as only existing in modern science fiction, but as you point out, it started in the seventeenth century, before there was any technology that could allow us to begin to think we could eventually do it. What compels us to imagine what was for a long time utterly impossible?

It didn't even take technology, and it didn't take the seventeenth century! Yesterday I mentioned the *Dream of Scipio*. For as long as people have been dreaming, imagining, and writing, they've been trying to see their earthly home from above, from its skies—even at times as if through a rear view window, escaping or leaving the planet behind. So I'm wondering today about Earth's gravity. Is gravity merely a physical force or is it also a pull that we can think about as an emotional draw as well? And if Earth has such double-pronged gravity, why is it that we're always trying to break free of it but then always looking back?

Gravity. There's that joke . . . gravity, it's not just a good idea, it's the law. We grow up and we don't notice gravity for a long time. Isn't it funny that when you become aware of it, sometimes you feel it so strongly and you want relief from it (perhaps that's just aging!)? Now we know what it's like to walk on the Moon, that you can take great bounds because the gravity is so much less. And it would be similar on Mars, more gravity than the Moon but so much less than the Earth. I think that has increased our wish to escape Earth's gravity.

But I think even in the days before it was realized to be a calculable force that objects and celestial bodies like the Earth and Moon exert upon us all the time, gravity was understood as an innate pull from the Earth that people could imagine as coming from individual stones as well as from the globe itself. Medieval people realized that this pull was physical and affective. Even stones were full of emotion-laden tug. And isn't it in part because of the Earth's gravity (in all the senses of that word) that we are writing this book together?

Yes, so right in all the senses of the word. I'm struck that you talk about space travelers looking back at the Earth, because I always think of them looking forward. For geologists and I hope for others the rocks on Mars have the same emotional allure of the rocks on the Earth, maybe even more so for their alienness. I imagine that this strange fundamental compulsion that people have had through the centuries to imagine leaving our planet was just enhanced by the discovery of long-distance balloon travel, and then by the discovery of rockets. We just kept taking the technological discoveries and cobbling them onto our existing dreams, making the dreams seem more feasible, more realistic. Now we can go to the movie theater and watch space travel so realistic that you feel like you're there yourself. The sensation of reality makes us think it *is* reality.

I like how you describe technology as enabling an intensification of human desires, wishes, and stories (things that are already there) rather than something that creates an absolute break with what comes before. Technology realizes more than disrupts. I actually think narratives are *real*—that is, they have a way of shaping facts and molding thoughts. Narratives can create possibilities (and worlds) substantially different from what would be ours if we didn't tell such stories or if these stories failed to take hold of our imaginations. There's a way in which Earth itself is always a story! Sure there's the physical object *Earth* that we study and want to know more about. Then there is the Earth that is a possible protagonist of a story that we are always working on. This Earth exerts its pull on and off throughout history, so that for millennia we have been trying to get to know the planet better. Over time we have carried these narratives along, sometimes without realizing it—I think that's why we call the Earth's primal eon the Hadean. And here we are again, another day of writing together about Earth, another approach to trying to understand its vast curving, and all I can think about is the giant globe in the lobby of this building where when you press a button you can see what the Earth looks like from the point of view of one of the numerous satellites circulating over it. That version of the Earth can be dizzying, a vision of circling the planet the wrong way—up and down from pole to pole, or at what seems a random, jarring angle. There's no totality to this Earth, just a conglomeration of satellite stories that invite you to see the planet as varied and variable, an image constituted from multiple points of view. It's been fun to watch children play with the device every day. It's been fun to experiment with it myself.

That beautiful great globe that will also project the surface maps of any of the well-mapped planets or moons from our solar system, and it is another way that our imagination and our technology has taken us beyond our actual reality. This notion that we can look at a picture of any object that we know the name of is a wonderful fiction. Space missions have photographed in tremendously high resolution the surfaces of bodies that before we could only imagine.

And so many people think, subconsciously at least, that they could find on the Internet a beautiful photo of any object that has a name. I ran into this problem when I was writing a series of books on the solar system and I wanted to drive home the notion that we only have pictures of the bodies that we visited with space missions through tremendous effort and sweat and tears and years and dollars. I wanted to show the highest resolution image that humankind had for every moon and planet. For even in some of the quite large moons of the outer planets we have only very rough images, highly pixelated blobs of blurry light with no details. I thought this was immensely instructional and would help to calibrate us about the many things in our universe that we have not yet seen. But my editors would not allow me to publish any images that were not above a certain resolution. They were perpetuating the notion that high resolution—the resolution that mimics our eyesight when we are actually present with the object—is the only reasonable depiction of reality, and that low resolution is an error, not just possibly the result of a technological miracle of a space mission that got no closer to that particular object.

That is so interesting. We need our blurry photographs to remind us to be humble, to remind us of what we don't know, what eludes us. We need to be able to acknowledge the uncertainty that remains within the objects that we are looking at— otherwise we are lying to ourselves. Familiarity is not understanding! We have been thinking together about what this book will look like in the hands of you, the person who is reading this transcript of our conversation right now. Yes, *you*. You're the person we are hoping that this conversation will include. We thought we might feature at some point in *Earth* a kind of flip-book series of images showing how the conceptualization of the planet as viewed from its exterior has changed over history: from medieval manuscript illustrations of what Scipio beheld (an Earth striped with polar, temperate, and torrid zones) to the *Blue Marble* and composite satellite images of today. We fear that given the house style and relative brevity of the Object Lessons series this flip-book progression might not happen . . . so if you beheld such images in a previous chapter, then we are really happy it worked out! But if your chapters lacked them, we want you to know that we thought that this book should contain not only words but also a series of earths that convey change and continuity and desire over time. We wanted you to have them, to think with them, to join a conversation that has already crossed centuries. There are many things that stay the same over time as the Earth is imagined from space: round immensity, participation in a system rather than solitude, disembodied perspective, attending to the smallness of the places inhabited by humans, the continued imagining of a point of view that no human being has until recently had the ability to occupy. The Earth from space is *always* beautiful. Artists spend a great deal of time crafting depictions at once precise and striking. They know how to capture the eye. Even in medieval manuscript illustrations you get what might be called a high-resolution visualization of the Earth, so that you can see as much detail as possible and luxuriate in its possibilities and be convinced that it conveys a true perspective, even if it is a work of art more than science.

> In my introduction to geology classes I have an artist come to class one day and teach us drawing. Then we walk around campus and draw pictures of the rocks used in landscaping or buildings. I am trying to help the students notice what they can actually see, and try to draw only that. This is outrageously difficult. If the student's mind does not comprehend what they see with their eyes they will draw either blank space or will fill in that space with something made up. And so our wily brains fill in what we cannot understand. This is one of the reasons I'm so excited about the opportunity to visit this metal

world Psyche that we're proposing to NASA for a mission. There are no images of Psyche that include anything filled in by the human imagination. The best optical image we have of Psyche is just a dot of light, as if it were a star. I would like all the children of the world to draw images of what they think a metal asteroid will look like, and then we'll get photographs of it, and we'll know. The solar system always surprises us. We always find things that we were unable to imagine before we went.

I want to say again how enthralled I am by this notion of a metal world. When you put it that way—Journey to a Metal World!—you offer an invitation to accompany you to a place that is *kind of* known (it's made of metal, it must be amazing, like a SciFi story) but then again can never really be known from an earthbound perspective. So we hope to send metal instruments there and find out what it's like on this protoplanet that could be a sibling of Earth. I'm haunted by the proposal that you put together for the Psyche mission. I've had so much fun sitting in your office flipping through the file and seeing how artists depict this object that has never been seen by human eyes. As you said, in the "real" images we have it's just a blob of light. I asked you last night how we know that Psyche is made of metal and you explained that it's partly from analyzing the spectrum of light bouncing off the asteroid as well as from its interactions with radiation. The truth of its composition is in these signals that we need technology to see. Yet the picture of Psyche that is included in your proposal resembles the Death Star from *Star Wars*. And there's something about a distant world that looks like it's a Death Star that's so completely fascinating. Who would not want to explore it? On the Internet that line from *Star Wars* gets repeated in many hilarious contexts: "That's no moon!" One of the reasons that line sticks in our minds is because we are fascinated by the possibility that humans might be able to create through art and technology a thing that looks like a celestial body but isn't, really. Or is it? Are we not hoping, and fearing, that we might become creators with the power to conjure up something only gods or astrophysical forces far beyond human power and scale have so far made? Next stop: a planet! Let's create an artificial Earth! Could we actually ever do that? Are we already doing that each time we draw or convey with words a radiant vision of the planet? An Earth that is beautiful and desired and perilous (because it exceeds us so vastly) exerts its gravity over us, and we cannot stop imagining it, holding it, painting it, giving it story, again and again.

That line, "That's no moon!" We say it over and over again in our science team as well. I've heard people say mournfully that technology has killed imagination, that the stark reality of scientific measurement, the unforgivingness of a fact, has killed imagination. And yet I think we've proven right here that technology has only enhanced imagination. We have the imagination all along, and then we incorporate the technology and the science and they don't fetter our imagination in any way: we keep imagining wildly with these new icons and these new vehicles incorporated into our dreams. Even the serious NASA engineers joke that our metal world Psyche is an enemy base.

Because how can they not? Something in celestial bodies offers a constant invitation to the imagination, to human creativity. Would we explore anything at all without that allure? We've been emphasizing throughout this book the ways in which the human imagination of Earth carries with it a very long tail: stories that we've been telling since ancient times, sometimes with the facts utterly wrong, sometimes with details or implications that have to be rethought or discarded, but what stays constant is a push to know more. We realize in the long run that we can never really know any object in its fullness, and certainly not the Earth. Yet we are obsessed with prodding and poking and trying to grasp more story. We are surface dwellers who aspire to sky-god views. Yesterday at the end of our writing session you talked about humans as apex predators. We imagine ourselves as residing at the very top of the world, looking down on everything as potential prey or material for our use (rightly or wrongly: in general this is what humans do). Do you think sometimes we could be a little more humble about our place in the world, and recognize that from that apex perspective it's not necessarily the most humane world we are imagining? I guess this is a long way of saying that we're writing a book about Earth for a series called Object Lessons. And because much of my training as a medievalist is in Latin, whenever I see the word *object* I think about where the word comes from: a verb [*ob+jacere*] that means *to throw in the way of*. So an object is both a thing in motion (it's in the state of being thrown, it's tumbling) and a stumbling block that can hit you with its force. An object interposes itself. An object gets in the way. It might make you fall to the ground and reconsider what you thought was your secure footing. It might make you realize the world is not so stable as you thought. An object is a cognitive challenge. An object can seem overly stable if we imagine we are at some apex looking down. But because we almost never know enough about an object like the Earth, it keeps tripping us up. Its gravity is relentless. We get pulled back. Or maybe (and here I am thinking about how spaceships work and Lindy correct me if I'm wrong) we can sometimes even use the gravity of an object to rocket past, curve all the way back around and then go shooting farther into space beyond, gaining multiple perspectives and speeds and possibilities along the way.

> Yes, you can slingshot around using the gravity of a planet or moon or the Sun, and you can accelerate, you can get a boost. I didn't know that about the origins of the word "object." They are so suitable! This entire book is a reflection of the ways that humankind constantly grapples with whatever is in front of it, with whatever object there is. There's always a little struggle going on. We are struggling to understand something. We are struggling to go somewhere. All of these grapplings, all of these struggles. I wonder if that is one of the reasons that exploration is so compelling, and space exploration is the ultimate exploration of all. Because only when you're at velocity, only when you're moving unimpeded toward your goal, do you temporarily lose that sense that you are struggling and grappling with your object.

But we all know what comes next because we have seen enough science fiction movies. You think you're going to get to the planet okay, you think you're weightless and your trajectory is fine and everything will be smooth . . . but technology is going to fail you, or some space particle is going to bombard you, or the surfacing of some unexpected story (sabotage! jealousy! capitalism!) is going to trigger a human drama in which you are taken out of your comfortable trajectory of movement and you are faced with the possibility of your own extinction. And all of this because you thought you could escape Earth and grapple with more distant things. I keep thinking about how the verb *grasp* means to hold something tight as well as to apprehend. For the most part we have a tough time cognitively embracing things without touching them or experiencing them—but not necessarily physically, we can grasp through images and words. Science fiction movies about travel away from Earth stage that challenge well. They are always about unexpected difficulties and the multiple ways in which we might be destroyed as part of a struggle that is both intellectual and embodied. We have to go through that peril to earn the joy that comes from exploration, from expanded understanding. Discovery often arrives with a bad name, because the word has been used to designate the "discovery" by the West of places that already had people in them who knew their lands well. Discovery has been used in the name of colonization, and historically carries with it deep unkindness toward human beings we have failed to call fellow. But I don't think that means we should abandon discovery, just that we should be more humane, more inclusive and more careful in the endeavors that we undertake. Nor should we ever try to leave the problems of the Earth behind as part of this process of travel. I like that as part of the conversation unfolding in this book about Earth we keep returning to the question of justice and what's at stake in trying to imagine Earth as shared space that might be apportioned more equally. I think neither of us wants to leave our fellow earthlings behind!

As you were speaking I was thinking about how you were taking us back from the huge, that is the planet and space, to the tiny. This big traveler that can be brought low by just a speck of space dust, a tiny bit of space trash that punctures their little bubble of oxygenated safety. And when you talked about colonization first, I thought how lovely that we don't have to control the violence of one group of humans invading the space of another group of humans when we travel into space, where there are no humans. But then I realized we were back to the tiny again, because we try so hard not to bring bacteria and fungus and the many microscopic denizens of the Earth with us when we travel. Those tiny beings are so persistent and so difficult to measure and we would rather that they didn't take root on another body that had no life before. And I was thinking about introducing bacteria to Mars, which we have certainly done with the missions that we sent despite our best efforts at sterilizing them—and so ironically there is life on Mars and it came from Earth—you then began to talk about justice on Earth. I thought that going back to the microscopic dust that lays low the space traveler, or the bit of bacteria that is now on Mars, seemed a bit, well, trivial. Because indeed it is thinking about the scale of the Earth and the grand scale of our humanness that we are hoping inspires all of us to choose the more optimistic of our many possible futures.

I like the idea of many possible futures. In a way there are many possible earths as well, and we have been exploring some of them as part of this project. Through this journey that we have taken together in letters, text chat, and physical conversations we keep zooming in and out, through the vast to the tiny and back to immensity again. Scale shifts really are what Earth provokes and needs in order for us to grasp some of what Earth does. Last night when we were talking about humans as apex predators I brought up the ways in which that notion immediately brings us back to bacteria. Humans can't live without a flourishing microbiome: we are an Earth to them. Meanwhile other bacteria are attacking our wounds, tripping us up, trying to bring us down from the apex that we imagine is ours alone. Paging H. G. Wells! Of course to the bacteria there is no apex, just proliferation and likely nothing more than that. Anyway it's interesting to think that life on Mars may well have been introduced by earthlings as we probed that planet. Maybe when we explore Psyche and other celestial bodies we will unintentionally introduce life into those places as well—and who knows what these creatures will do when no longer earthbound. Life exceeds us at every scale, thriving without our even noticing, most of the time. Like Earth. Life and Earth both boggle the mind . . . and cannot be thought of without each other's intimacy.

7 INTERLUDE: A HIKE AROUND PIESTEWA PEAK

Warm afternoon in early December with mixed clouds and sun. Jeffrey Jerome Cohen, Lindy Elkins-Tanton, and James Tanton hiking. An edited transcript of a conversation recorded by iPhone.

So if somebody comes to this place to see a flash flood begin, they must know they're playing with danger. Do they lack enough imagination to realize what's going to unfold when the flood comes tearing through, and that lack's going to possibly kill them, or is it that they just want the adventure of being present?

Oh that's interesting! It isn't enough to simply imagine— even though we just imagined the flood, we all looked at the telltale signs of how often this basin gets swept by water. Imagination brings us here, but maybe we know reality is much more exciting than anything we can bring in our minds. Something about the love of the adrenaline, the danger. . .?

Or we know we will see things that we do not expect to see. Like that a flash flood begins with a muddy trickle, not a gush. When you witness that, that's foreboding. Not a deluge but a seeping. Then the flood.

And you record that rush because you have to share the story afterwards, through words or videos. That's key. YouTube is a hearth, you post your video there to make a community. Experience and imagination.

So did we decide that imagination is always bound up in visualization, and creativity is an act of making things, hoping they will become facts?

And that facts are made, so that even though they sometimes come apart again, we have to trust they will be solid and part of a foundational bedrock?

What's the difference between an act of imagination and or creativity and an actual solid fact?

Or how do we know that difference? That is the problem. You *hope* everything that you try to prove or that you stake your reputation on or that you publish is a fact. You really can't be confident. It takes years for people to be convinced by it, or to repeat your experiment, or even to read your paper. Or book.

That's just it. What makes a fact? Is it repeatability?

Well, that's what people like to say about science, but not all science is repeatable. Geological science, planetary science, much of it is not repeatable. The scales are too vast.

So the Earth is imaginary and it's not exactly a fact? (laughs)

Do we like this *Earth* book still? Do we like where it's bringing us?

No, I actually agree with that! Earth's imaginary and not a fact. Or at least, I can't insist, that it's not imaginary and it's not a fact. There are certain assumptions that I'm not giving up on, but I could be wrong. Earth's imaginary *and* a fact?

So I suppose that here is where repeatability comes in. I think it's kind of silly to say that the Earth is imaginary and not a fact. By that what I mean is, the only thing we can attest to is what we can see and feel right in front of us. So we have a certain horizon all the way around. We can attest to the top of that hill, and that blue mountain in the far distance, and this whole part of the town, but not to the part of the town that we *think* is still on the other side of the mountain behind us. But repeatability tells us that all the other people on Earth, or some majority of them, experience and see their parts of the Earth all at the same time all the same way, they describe them the same way, all the pictures come back day after day the same, and so we have to say the Earth is a fact.

So a fact is something that *works*, and often something we have made together. . . . A horizon is a thing we try to rise above, to see the world from its outside, hoping we'll glimpse something stable. Assemble some separated facts into larger pictures. Escape the surface and the small horizon, get a bigger image, maybe the whole globe. The *Dream of Scipio* is alive and well!

But it takes someone, or a group of someones, to push at that horizon, to go somewhere outside the current facts.

Re-center the Earth. Or de-center it. Explore.

So is a fact really just that many people agree on how it looks, feels, and responds?

Well, welcome to mathematics!

I think that's really what it is, isn't it? And so repeatability can be, in the strict double-blind data sense way, or as agreed upon by multiple people through their own senses.

Facts have to be put together and they seem to require a fair amount of belief to work well in the world. There is an element of faith behind many facts. And that unthought belief can be an impediment.

And maybe that is yet another reason why some people senior in their careers tend to protect their turf so forcefully? It's not just that they don't want the next people coming with the next idea to overturn their idea, it's that that *is* their reality. They have thought within it for so long it's part of their identity and their sense of how the universe is and so it is disturbing on a much more fundamental level.

But does it have to be disturbing?

Well, it depends upon how relaxed you are!

I pretty much figure it happens to you as a scholar or writer no matter what. Somebody is going to come along with something better and more exciting. That's OK, we share a field and there is no point in being territorial about it and let's make the field as exciting as possible. It's not about the longevity of your primal idea—it's about the conversation.

Oh, but see, I know a whole bunch of people who will tell you explicitly that that *is* what it is about. I know people who say to me, my goal is to answer something so that in a thousand years people will know me as the person who discovered x.

They wrote the book on it. The book stays closed!

"I want to be Newton, I want to be Euler!"

I guess it is a nice aspiration but I don't see how it can be true for the majority of scholars.

No, of course not, but within a certain cadre, there is the feeling that this is the cadre that someone in it is going to be that person, so why not me?

And that is why they are so evil in peer review!

That's true, that can be a part of it.

Territorialism. As if the Earth weren't big enough. Small regions claimed and owned.

And that's why we need more spaces for challenging conversations to unfold, more unexpected gatherings, more engagement and communities.

More horizons to explore. More imagining and creativity. Till we push too far!

There is a phrase in mathematics which is "don't stare too deeply into the abyss, lest you go mad," which is a thing about infinite numbers.

Can you die from an overactive imagination?

Oh, like that thing if you die in your dream you die in real life?

Well, I think people say that because they are afraid of their own imagination, of their own ability to push their horizon, to the point that they imagine they will perish if they plunge themselves into the unknown.

I do think that over a long term a person doing that can drive themselves to the brink of something . . . some disorder . . . do you think so?

An overactive imagination can lead you to some very, very bad choices and very bad situations.

We had to imagine the Earth in the distance before we could achieve the view. And be surprised by it. Without overactive imaginations, where would we be?

Well, here we are, relatively far out into the desert.

8 IMAGINATION

Dear Lindy,

I am safely back in Washington, DC, after a wonderful sojourn to Arizona. Thanks so much to you and James for your hospitality. I already miss the morning lattes and evening prosecco, the walks to see Christmas lights in the desert, the nightly digestifs of amaro, the conversations about everything from the challenges of family and making the life you want to imagining the fate of the universe. My brain is still on overload. But it was wonderful to be a member of your household for a week . . . and to be part of the excitement at Arizona State University as well. I was truly impressed by the School of Earth and Space Exploration. From the moment you walk into the main building, you are filled with wonder: a model of the Mars Rover to one side, open labs everywhere, a projection globe in the middle of the lobby, an impressive collection of meteorites nearby. The school's name seems apt, bringing together Earth and space as shared endeavors of discovery. As was quickly clear to me as we embarked on this project, exploring space means exploring Earth and vice versa: we leave home to know it better.

We ended our time together with a hike around Piestewa Peak. As you, James, and I wandered through the arid landscape I kept thinking about how much the gray, red, and green stone looked like the terrain of a distant planet, and your remark in a letter last autumn that compared to a place like Mars (or even to many deserts on Earth), this expanse of Arizona is saturated—something like seven inches of water per year. The alluvial fans revealed a rich story about how water shapes the Earth, while the rumpled peaks that dot the reserve suggested that over long periods of time the surface of the planet undulates like the sea. As we followed the rocky trail we debated topics like: Why do people film and share natural catastrophes? Can you think yourself outside of yourself? Outside of the Earth? What's a better assessment of a college education's impact, maximum wage or maximum life? How long do accidental human and animal impressions on terrain remain readable? Are wild quail really all that wild? Must results be repeatable to be believable? What about unrepeatable experiments, or durations of time that cannot be captured by human-scale inquiry? Where is the edge between knowing and imagining? Given that facts are made, aren't they always in part imagined? Do facts impede when we take them *as* facts? Should we embrace the likelihood that any scholarship we do will be outgrown? Should we forgive evil peer reviewers (and others nasty to us as we try to do our work, live our lives)? Is the Earth as we know it largely imaginary? Can you die from your imagination?

Heady stuff. Today as I sit in a coffee shop before a series of meetings that will propel me back into mundane academic life, I am struck by how frequently we returned to the role of creativity in art, criticism, science, and mathematics. If facts are things that are at once discovered and made (meaning that many of the facts we know fall apart over time, and it can be difficult to discern the difference between things that are true and things that are believed to be true), then exploration is a rather open-ended process that requires, for success, a willingness to challenge, a compulsion to think beyond pregiven limits. At the ASU holiday party, one of the deans spoke about a brilliant dissertation that posed a question that to many experts had seemed ridiculous: might DNA be exchanged not just across species but across kingdoms? No one wanted to sponsor that graduate student's inquiry, so finding the lab space to conduct research proved a great challenge. Persistence paid off though, and the student revealed something about how the world works that we otherwise would not have known. I think about this kind of tenacity or drive or obduracy a great deal: What is it that compels some people to ask absurd, impossible questions and to refuse to yield when told they should find a different path? Most who do so probably fail and we hear nothing about them. And that must be the key: a willingness to risk foundering, defeat, even humiliation—and since we are talking about exploration, even death. Think about how many would-be discoverers set off for unknown islands only to be swallowed by the sea. Creativity means bringing something new into being, knowledge or technology or art—most likely all three of those at once. Creativity requires the ability to persuade yourself, first, in the hope that what you find or make will persuade others. It's never as solitary as it seems. Even when we are in our own heads processing ideas, we are always collaborating with others. We are always supported by networks, even if fragile ones. We are never alone in the desert.

OK, we sometimes are, but that's an easy way to perish. I do believe that imagination can be lethal. Yet even if creativity often seems a trait of individual genius—that's the way history is taught, after all—none of us work completely alone. This book would not have been possible without legions. The Earth as I have come to know it over the past few years is a different planet from the one on which I had long dwelled. More fragile, certainly: what I have learned about the scarcity of its water and the thinness of its atmosphere makes life's precarity palpable in a way that it wasn't before we embarked on this project. Despite remaining too vast for me as an individual to comprehend, the Earth as an object keeps oscillating between imagination and hard fact, between idea and geological reality. It's in that wavering uncertainty that its power over me seems so tangible. I admit, in the end, that I am writing this letter and this book because I want to create with Earth. And with you. And with the stories we earthlings have long been telling about our home.

We've often spoken during our ongoing conversation about what imagining the Earth as a beautiful object suspended in space achieves. Many writers and thinkers believe that when viewed from afar, our luminous planet provides the perspective we humans need in order to become a little more humble about our small place upon its vast surface—and maybe even to realize that we are a community of Earth-dwellers, united by our shared home. Apollo 17 gave us *The Blue Marble*, the most reproduced image in the history of images, and a catalyst to 1970s environmentalism. But then I think about Scipio razing Carthage shortly after he dreamed the Earth from space and came to know the tininess of the Roman Empire. I also think about the words of Apollo 9 astronaut Russell Schweikart, who beheld the Earth during a spacewalk and was overcome with awe. In 1969 he asked some questions that still haunt:

> There you are. Hundreds of people killing each other over some imaginary line that you're not even aware of, that you can't see. And from where you see it, the thing is a whole, and it's so beautiful. You wish you could take a person in each hand, and say, "Look. Look at it from this perspective. Look at that. What's important?" (Robert Poole, *Earthrise: How Man First Saw the Earth*, 165–66)[9]

Yet despite the shared perspective this Earth from space summons us to inhabit, few of us seem able to accept the invitation, at least not for long. The lines between nations might be revealed as imaginary from space, Rome and Troy and Carthage might shrink to dots, and yet war continues unabated. Individual, municipal, and national interests take precedence over global concerns. Resources continue to be squandered. The resplendent and fragile Earth as Blue Marble activates the imagination for a little while, perhaps, but does not necessarily prompt much action. We remain earthbound.

Why when it comes to Earth does human imagination luxuriate in beauty but not initiate action and change? How might we do something with and for this Earth that we are at last able to behold from a distant perspective? Anand Pandian has recently written that the salvific promise of the *Blue Marble* image has proven an illusion even as the problem of perspective it embodies has become more urgent.[10] Now that we know we have entered the Anthropocene, when human industry significantly interacts with and alters the complicated climatological systems of the Earth, what do we do to ensure that this planetary home remains habitable for its human dwellers (and not just for some but for all)? Or do we simply admit that all things have lifespans, even species like *homo sapiens*, and that the Earth and Life will thrive even in our absence, even if in ways we simply cannot foresee? Pandian asks: "Do we need more of the lucid vision of daylight or the imaginative powers of some other kind of dream? What will it take to nurture and sustain this kind of vision?" ("Seeing Things"). Daylight and dream: science meets art? Data meets story? Can imagination keep us from despair?

—Jeffrey

Dear Jeffrey,

Today is Christmas day, and we are in our little house in western Massachusetts. In a few days our son Turner and his girlfriend Liz will join us. My brother and his wife are spending the holidays in Ireland and Italy, visiting family, after the sad death of her mother. So, for a few days, it's just James and me.

This landscape is so deeply familiar to me, and so different from Arizona. It's been bizarrely warm, almost 70°F today, and not a flake of snow has been seen, even here up in the hills. The warmth keeps the grass reluctantly a bit green. We have seeded our meadow with native bunch grasses, and they throw up sprays of stems six feet tall and capped with exuberant seed heads, in many shades of gold and russet. Every part of the earth is covered with plants, so different than our Arizona desert.

We walked down our long drive. Halfway to our creek we found, as we do almost every time on the morning after our arrival, almost every time, a large coyote scat placed meaningfully in the center of the gravel driveway. We know the land belongs to you, coyote, and we promise not to eat your food.

We met our neighbors a tenth of a mile down the road, and together we walked another half mile up through the woods to see a new historical marker they were eager to show us. How different from our hike in the Arizona desert just a couple of weeks ago! This evening I listened to the recording I had made of that walk, and through it all was the loud crunching of our boots on dry scrabbly rocks. Today, we had the soft slippery sound of leaf mold, and the smell of mushrooms, and some cries from ravens up in the old maple trees.

Those maple trees, 200 years or so old, are, I think, the remnants of a movement by Quakers, espoused by Thomas Jefferson, to weaken the immoral slave trade by manufacturing native maple sugar rather than buying cane sugar from the Caribbean. These old maples, all about the same age, line old roads throughout New England and they are all at the ends of their lives. They are literally falling apart, great limbs crashing down, huge hollow trunks standing in the forest, so much larger and grayer and more severe than all the other trees.

Halfway up an old cart road alongside Chapel Brook we came to the historical marker. Its title reads, "Site of the Alvan Clark Homestead." The nearby rectangle of collapsing stone barriers around a sunken area, originally the walls of the basement and now inhabited and surrounded by hundred-year-old trees, are all that remain of the homestead the Clarks had. Alvan Clark and his sons, the marker goes on to say, became world-renowned experts in grinding and polishing lenses for telescopes. Their lenses were used all over the world, including in Arizona in the Lowell Observatory's legendary Clark Refractor, the telescope Percival Lowell originally used to search for signs of life on Mars.

Clark went on to get honorary degrees from Amherst, Chicago, Princeton, and Harvard, but his only formal schooling was at the one-room schoolhouse just up the cart path. Like his family's homestead, the schoolhouse exists now only as a tumbled stone foundation in the forest. We stood in this forest, a hundred and more years old, and imagined the past. When the Clarks lived on Chapel Brook, the land was cleared everywhere in New England. The earth was tilled and run with livestock. We squinted through the gray winter trees and tried to imagine the shape of the land without the forest.

And thus I was connected in so many ways to our conversation while walking in the desert. There was Clark, on a prosperous farm and mill community right down the street from our current day wild forest with bears and porcupines, making these telescope lenses. The telescopes he helped build were the state-of-the-art of their day, enabling scientists to make observations that were never before possible, enabling the solar system to surprise them and present them with sights that surpassed their imaginations, just as our space missions do for us now.

Where does the known end and the imagination begin? For centuries people have imagined they could go away from the Earth and look back at it, and they imagined that visual. For a long time people just imagined it, but they couldn't do it. They would look around them on the ground and imagine what that might be from another perspective that they have never had. And so they have some little fact to begin with: here is what I see in front of me. Their image of the Earth seen from above contained that little fact, plus things they had heard about other parts of the Earth, plus some more pure imagination.

And now, we can see a picture from a space mission of what the Earth looks like from above. So I am wondering where the edge is between knowing and imagining. That ragged area is where science sometimes gets into trouble, and where it's difficult for people outside of science to understand why something that we think is right can turn out to be wrong. And it's not just because of simple things like "I made the measurement wrong." The greater danger, or opportunity, is that sometimes ideas are so potent in our minds, become so realistic in our imaginations, that we take these ideas more for reality than we should. I can describe a model, for example, for what a baby planet, a planetesimal, very early in the solar system, might have looked like. I think it would look this way because of the following models or physical ideas. And then that image can become so energized and vibrant in the minds of other researchers that they kind of take it as fact and build upon it, even though parts of it might be really wrong. Thus the imagination we need for science can cause us to make errors, but it also allows us to make progress.

There is a stereotype that science is formulaic and repetitive and narrow, and scientists sit and make measurements over and over and over again and then make a graph, or are just calculating, calculating, calculating. But in the natural sciences there are many moments when you need to be extremely creative to make progress. One is in framing your question, which would seem to be simple, but to ask a good question you have to have an idea about what is happening.

And to have an idea about what is happening you need to have a real imagination about it. You have to imagine the length scale and the time scale and the temperature scale, and you have imagine all the things about how the parts of the process might fit together, until you have an idea about how it might be working. And then after you make whatever measurements or model or calculations you can do, then you have to have a wonderful imagination to think of the different possible things that your data could mean, and figure out how to discriminate among them. So there is a need for a fantastic imagination, where you can picture processes running like filmstrips, or get an intuition for the material and imagine how it might react in different circumstances. It is like making up a story in your head, and then comparing that story to the world.

Jeffrey, you wrote, "Creativity requires the ability to persuade yourself, first, in the hope that what you find or make will persuade others. It's never as solitary as it seems." Recently I have been talking with a friend about leadership. He has led all of his life, and he has taught others to lead, often with the aim of succeeding in business. He explained to me that leadership is the act of having a vision, believing in it, and pursuing it. Other people may then follow. Thus, creativity, vision, and leadership are all bound together; your statement about creativity is the same as our statement about vision.

Your next question "Why when it comes to Earth does human imagination luxuriate in beauty but not initiate action and change?" then becomes answerable. There must be a person who uses their creativity to produce a vision for change, and then they must lead toward this vision in a way that others can follow.

There are two beautiful things I have learned in writing this book. One is that the chasm between humanities and sciences is an optical illusion. We are all using our creative brains to interrogate the world, and any differences in our perspectives or techniques are completely overwhelmed by the vast commonalty of our humanness.

The second is that humans are searchers, seeking to understand truths about the world around us. Alvan Clark was seeking to understand the universe (he even published papers about stars, later in his life). Our understanding of the universe is constantly changing, though. As scientists say, almost everything that we ever think that we know today will be proven wrong—it's scientific progress. And first you think, oh, that's so lovely, so humble, and you think about progress, and then you think for another second, and you say, then *what's true*? If you ask scientists what is really a fact, it is difficult to find more than a handful of things that we think we will agree on forever, immutable truths we have found.

On our walk in Arizona we were talking about the acts of imagination or creativity that bring you to something you hope will be a fact. I asked you, what is the difference between the thing you hope will be a fact, and a fact? And you said, in the end, a fact is something that many people agree on.

We hope they agree on the basis of careful observations and thought, and not on emotional reactions. Emotional reactions create weak opinions. Care and thought and practice in knowing things creates more lasting kinds of knowledge.

So let us, and let all who feel able, both luxuriate in beauty *and* initiate action and change. We each can find a vision to lead us to an optimistic future, and we can lead with our visions. Let's start.

—Lindy

A Facebook status update that became a blog post

December 26

So Lindy sent the last letter in our *Earth* book to me last night, December 25, a festival of light and life against winter's chill.

We don't celebrate Christmas, but my family loves the various traditions that cluster around the solstice: candles, food, and merriment when cold nights grow long. One of my favorite stories for this time of year is the medieval poem *Sir Gawain and the Green Knight*. The poem affirms seasonal vibrancy (green holly and red berries, feasts and warm fires) without disregarding winter's cold violence or things that exceed small human frames (red is also the hue of flowing blood; animals suffer and shelter along with humans; green throughout the poem is supernatural in its ability to stun, challenge, and stealthily thrive). Unlike the contemporary poet Geoffrey Chaucer, whose *Troilus and Criseyde* exults in a moment of viewing the Earth at great distance (so that it becomes "this little spot of earth, that is embraced with the sea"—Chaucer was a fan of the *Dream of Scipio*!), the anonymous author of *Sir Gawain and the Green Knight* never gives you a moment of rising above terrestrial entanglements. He is never tempted to imagine a view of the planet in its entirety, as some distant orb, since that view would diminish life among the earthbound. Sleet and wind-driven ice batter knights, horses, and shivering birds alike, just as the Sun's warmth delights even plants. It's hard to take an easy moral from the story, especially because the Green Knight is as full of vitality as violence, peril entwined with exuberance, red and green. With his great ax and his ability to survive his own death, he's a fearsome creature. But he's also a jolly companion who will forgive you for lying and decide at the last minute not to chop off your head, then invite you back to his castle for cocktails and a New Year's party. Facebook yesterday reminded me of the medieval poem's intertwining of themes because so many friends posted about the Krampus (the horned and hairy devil who punishes naughty children) and Jólakötturinn (the Yule Cat who devours those who do not leave offerings), along with pictures of garlanded fir trees, gifts torn open by the eager young, and plentiful cakes. It's traditional to tell ghost stories around the winter holidays. Maybe the Smiths said it best: "In the midst of life we are in death, etc." Can we have an un-ironic version of that?

Precariousness is also on my mind because we just made our annual December return to New England to see family and celebrate my dad's birthday (his eighty-fifth). At various family gatherings stories were retold about how many times things went badly wrong, and how persistence and good humor often enabled recovery. When he arrived in Maine in 1882, my great-grandfather, an immigrant from Lithuania, made his peddler's way from farm to lonely farm in Penobscot County. He was for many Yankees the first Jew they ever met. He eventually saved enough money to open a shop in Bangor, then a chain of clothing stores across the state, but lost everything in the Depression. Sudden turns of fortune, the unkindness of family toward family, and eventual peace are recurring themes of these yearly stories we tell. I am thinking about all this because Lindy's letter contains a poignant meditation upon houses reduced to ruin and encountering human history as it vanishes into landscape. Few of my relatives now remember that my great-grandfather's name was Simon, and fewer still know that it was really Shimson. Today you will not find many traces of the once lively Jewish community in Bangor. But you will find something, if you look with enough attention.

When Lindy sent her letter I was on an airplane back to DC, descending through so much night rain that it seemed we were on a ship with battered portals. Alex is just back from his first semester in college. Katherine has completed about half of her first year of middle school. Wendy continues working two jobs well, the second as a local elected official.

Sometimes I think that time is propelling the four of us so quickly forward (how is it that we now have an eighteen- and an eleven-year-old?) that it's like we're always on that plane, onward relentlessly toward destinations we can't clearly see, trusting we will safely arrive. We landed at BWI, happy to be under the storm rather than within it. As we taxied for the gate I checked my phone for email. I read Lindy's letter while we waited for delayed luggage and as we took a shuttle bus to our car. Its close is so full of hope and promise that I knew it had to end the book, even though she and I didn't plan it that way. That sudden realization surprised me with the pang of sadness it brought. I do not want the conversation to end.

Earth is a problem. In my last letter I had worried that awe for its beauty can lead to political and ethical paralysis. Too often people convince themselves that it's enough to praise the splendor of the planet in its solitude: the Blue Marble as talisman. Imagination propels us to find new modes of comprehension but it sometimes immobilizes or betrays. How do we ensure that appreciation yields to endeavor? Lindy wrote (and I hope she will not mind my repeating her words here, but they seem so right as we approach the New Year):

So let us, and let all who feel able, both luxuriate in beauty *and* initiate action and change. We each can find a vision to lead us to an optimistic future, and we can lead with our visions. Let's start.

This book comes to its close as yet another rotation of the planet round its warming star completes—a cosmically insignificant fact that means the world to us earthbound observers, who need to pick *some* place to start and to end, and then to begin again. A lure for the imagination, a catalyst to creativity, and (if we are lucky) a spur to vision and engagement, Earth is too vast to be encompassed, especially in a book so small. Earth is a shared project, beautiful and incomplete.

Yes, let's start.

ACKNOWLEDGMENTS

We thank Christopher Schaberg for inviting and enthusiastically supporting this little book about an impossibly large object; Ian Bogost for challenges to frame the project better; Haaris Naqvi for never saying no; and Susan Clements for supportive shaping of what you now read.

Jeffrey Cohen is grateful to his family (Wendy, Alex, and Katherine always challenge him to ask the biggest questions wherever we roam); his siblings, friends, and colleagues; the George Washington University for research support; and James and Lindy for opening their home, opening their lives, and opening his mind beyond anything the two little neurons he possesses could otherwise accomplish. He cannot imagine a better collaborator in all things Big than Lindy Elkins-Tanton and is so very happy to share this Earth with her.

Lindy Elkins-Tanton is grateful every moment for her family, James and Turner, and their never-flagging eagerness to discuss the many fascinating questions in this universe we share; for Jimmy and Margaret, and our endless adventures; for

Arizona State University and my wonderful colleagues there and our sense of teamwork and promise (there cannot be a better place for an interdisciplinary explorer), and especially for Jeffrey Cohen and his exceptional collaborations. I have never had more fun and discovered more new ideas than when writing with Jeffrey. So thank you, Arthur Bahr, for introducing us and making us do that plenary conversation together those many years ago!

LIST OF ILLUSTRATIONS

NOTES

Chapter 3

1 Geoffrey Chaucer, *The Riverside Chaucer.* Gen. ed. Larry D. Benson, 3rd ed. (New York: Houghton Mifflin, 1987).

2 Charles Darwin, *The Origin of Species by Means of Natural Selection, or the Preservation of Favoured Races in the Struggle for Life* (London: John Murray, 1859).

Chapter 4

3 Ursula K. Heise, *Sense of Place and Sense of Planet: The Environmental Imagination of the Global* (Oxford: Oxford University Press, 2008).

4 Henry of Huntingdon, *Historia Anglorum: The History of the English People*, ed. and trans. Diana Greenway (Oxford: Clarendon Press, 1996).

5 John Mandeville, *The Book of Marvels and Travels*, ed. and trans. Anthony Bale (Oxford: Oxford University Press, 2012).

6 Anand Pandian, "Seeing Things." *Public Books* (December 1, 2015), http://www.publicbooks.org/nonfiction/seeing-things.

7 Robert Poole, *Earthrise: How Man First Saw the Earth* (New Haven: Yale University Press, 2008)

8 Theodor Schwenk, *Sensitive Chaos: The Creation of Flowing Forms in Water and Air Revised Edition*, trans. J. Collins (London: Rudolf Steiner Press, 1996).

Chapter 8

9 Alan Weisman, *The World Without Us* (New York: St Martin's Press, 2008).

10 Jan Zalasiewicz, *The Planet in a Pebble: A Journey into Earth's Deep History* (Oxford: Oxford University Press, 2010).

INDEX

Page references for illustrations appear in **bold** type